Motivating A-level Mathematics

MOTIVATING A-LEVEL MATHEMATICS

A Source Book

BY

THE SPODE GROUP

Oxford   New York   Tokyo
OXFORD UNIVERSITY PRESS

Oxford University Press, Walton Street, Oxford OX2 6DP
Oxford New York Toronto
Delhi Bombay Calcutta Madras Karachi
Petaling Jaya Singapore Hong Kong Tokyo
Nairobi Dar es Salaam Cape Town
Melbourne Auckland
and associated companies in
Berlin Ibadan

Oxford is a trade mark of Oxford University Press

Published in the United States
by Oxford University Press, New York

© *The Spode Group, 1986*

First published 1986
Reprinted 1989

Individual pages of this book may be reproduced
by individual teachers for class use
without the permission from the publisher.
With this exception, the material remains copyright
and no part of this publication may be reproduced
without the prior permission of Oxford University Press

British Library Cataloguing in Publication Data
Motivating A-level mathematics: a source book.
1. Mathematics 1961
I. Spode Group
510   QA39.2
ISBN 0–19–853653–4

Library of Congress Cataloging in Publication Data
Main entry under title:
Motivating A-level mathematics.
1. Mathematics—Study and teaching (Secondary)
2. Mathematics—Study and teaching (Higher)
3. Motivation in education.   I. Spode Group.
QA11.M66 1985   510   85–25830
ISBN 0–19–853653–4 (pbk.)

Printed in Great Britain by
St Edmundsbury Press,
Bury St Edmunds, Suffolk

Contents

List of contributors	vii
Introduction	ix

1. Functions

1.1. Oscillations in nature	3
1.2. Party representation and election results	6
1.3. Growth rates	9
1.4. Apparent magnitudes of stars	11
1.5. Shortest paths across rectangular cities	12
1.6. Gravitational force	15
1.7. Investment and borrowing	18
1.8. Scree slopes	21

2. Calculus

2.1. The shape of a tin can	25
2.2. Fleet size for car leasing company	27
2.3. Stock control	30
2.4. Sales response to advertising	33
2.5. Radioactive decay	37
2.6. Carbon dating	40
2.7. Art forgeries	42
2.8. Drug absorption	45
2.9. Population models	50

3. Mechanics

3.1. Pursuit curves	61
3.2. Modelling river flow	65
3.3. The tennis service	70
3.4. Head-on-crash	74
3.5. Braking a car	79
3.6. Experiments in mechanics	83
3.6.1. Falling freely	83
3.6.2. Throwing surprises	84
3.6.3. The speed of an oscillator	85
3.7. Kepler's law	87
3.8. Industrial location	90

4. Probability and statistics

4.1. Are you being served?	95
4.2. Quality control	97
4.3. Blood donors	100
4.4. Intelligence quotients (IQ)	102

Solutions to the exercises
    1. Functions   107
    2. Calculus   112
    3. Mechanics   119
    4. Probability and statistics   126

Contributors

John Berry	Faculty of Mathematics, The Open University (Editor)
Roger Biddlecombe	Ounsdale High School, Staffordshire
Roger Blackford	Computing Advisory Teacher, Staffordshire
Brian Bolt	School of Education, University of Exeter
David Burghes	School of Education, University of Exeter (Editor)
Bob Davison	School of Mathematics, Leicester Polytechnic
Nigel Green	Tong School, Bradford
Ron Haydock	Mathematics Department, Matlock College
David Hobbs	School of Education, University of Exeter
Ian Huntley	Department of Math Sciences, Sheffield City Polytechnic (Editor)
Robin Ingledew	Stainsby School, Middlesbrough
Peter Moody	Mathematics Adviser, Dyfed
Graham Nellist	Boynton School, Middlesbrough
John Walton	Monks Walk School, Welwyn Garden City
Barbara Young	Tarporley County High School, Cheshire

The Spode Group

Director	Professor David Burghes
Associate Directors	Dr John Berry and Dr Ian Huntley
Secretary	Ms Sally Williams, School of Education, University of Exeter, St. Luke.s Exeter EX1 2LU, UK

Introduction

This book illustrates many applications of topics in A-level mathematics. The range of application is not comprehensive, but we have tried to include what we regard as some of the most important applications, whilst others have been included as we think that they are basically interesting problems.

We have divided the book into four chapters

1. Functions;
2. Calculus;
3. Mechanics;
4. Probability and statistics.

Each chapter contains a number of case studies. Each case study is divided into an introductory problem statement, solution, and further exercises, and the mathematical topics needed for the solutions are stated at the beginning of each chapter. The book can be used in several different ways. Each case study could be used by A-level teachers as a motivating example to start a particular mathematical topic. Alternatively, the case studies (and exercises) can be used after a mathematical topic has been taught in order to reinforce ideas and concepts and provide an insight into the practical applications of the mathematical topic under study. A third way is to pose the problem in the form of the first section of each case study and let students attempt to solve the problem with or without guidance. These alternative strategies range from descriptions of mathematical models to modelling exercises depending on the teaching mode used. We believe that all these strategies are important in the teaching of mathematics.

Many of the models presented are well known and have been used in practical situations (e.g. 1.6, 2.6, 2.7, 3.5). Other models are more speculative in that they explain what is happening rather than being precise models that can be used to make predictions (e.g. 1.2, 2.9, 3.2). A number of the models presented in Chapter 3 can also be found, together with additional material, in another publication of the Spode Group called *Realistic applications in mechanics*.

It is vital that A-level students, as well as being competent at using mathematical routines, should also see their applications and have some experience in applying mathematical topics to practical problems. We hope that this book will provide both student and teacher alike with a valuable resource that can be dipped into at various stages throughout the A-level curriculum.

The writing of the first draft of this book was financed by a grant from the Schools Council. We are grateful for their help and particularly for the enthusiasm of their mathematics curriculum officer, Mrs Jasmine Denyer. We are also grateful to the I.M.A. and, in particular, to Norman Clarke and Catherine Richards for their encouragement and to their consultant, Professor Geoffrey Matthews, for his detailed comments which have turned our first draft into its present form.

1. Functions

In this chapter we present a collection of problems showing examples of the *use of functions* in providing models for certain situations. It is the manipulation and application of these functions that we are illustrating, not the physical situation itself.

1.1. Oscillations in nature

- **Sine function**

Birds and insects are able to fly by a complicated process of flapping their wings. The graph in Fig. 1.1 represents the *oscillations* of the wings of a locust in flight. The solid line represents the motion of the hindwings and the dotted line represents the motion of the forewings. Graphs of this type are common in nature where we have an oscillatory type motion. They represent functions that are called *periodic*. The feature of a periodic function is that at regular time intervals, T say, the function has the same value. T is called the *period* of the oscillatory motion. For the motion of the locust's wings shown in the diagram, the period is almost 0.06 seconds.

Find a formula in terms of sine or cosine functions to model these oscillations.

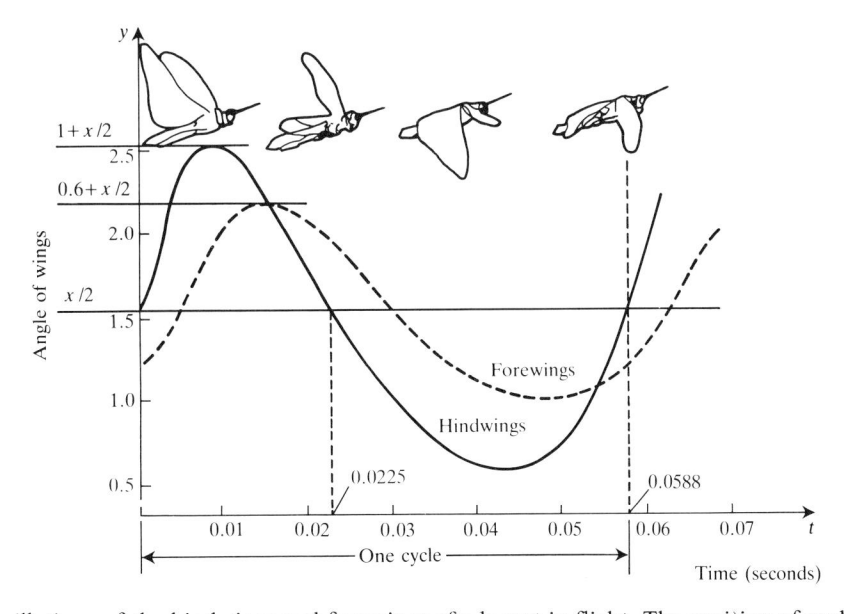

Fig. 1.1. The oscillations of the hindwings and forewings of a locust in flight. The position of each wing is measured by the angle y between the actual wing position and the downward vertical position; these angles are plotted as a function of time for each pair of wings. At the starting time $t = 0$ the hindwings are halfway up, at an angle of $x/2$ with the vertical. The forewings oscillate slightly out of phase with the hindwings. [After T. Weis-Fogh (1956) 'On the flight performance of locusts.' *Phil. Trans. R. Soc., London, Ser. B.* **239**, 494.]

Functions

- **Solution**

The trigonometric functions, $\sin \omega t$ and $\cos \omega t$, exhibit the same features as the graph of the motion of the locust's wings. They are periodic functions of period $2\pi/\omega$.

Let us use the trigonometric function $\sin \omega t$ to model the motion of the wing.

Consider first, the movement of the locust's hindwings. Their motion is periodic about the horizontal. The period of the motion is approximately 0.06 seconds. The largest angle that the wings make with the horizontal is approximately 1 radian and this is called the *amplitude*.

The function $\theta = \sin(2\pi t/0.06)$ exhibits the same features as the solid line in the figure. This function does not fit the graph exactly however, (see Fig. 1.2), and an improvement would be to find a sum of several different trigonometric functions.

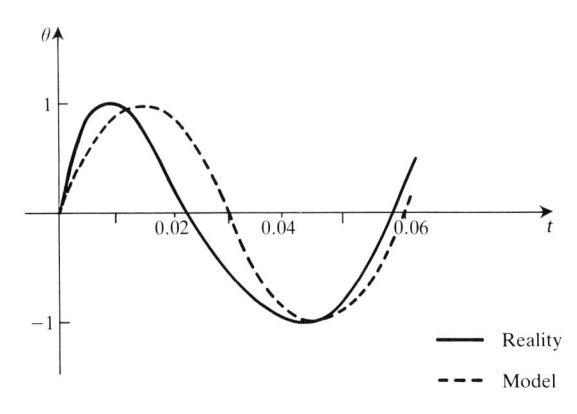

Fig. 1.2

Now let us look at the motion of the forewings. They lag slightly behind the hindwings. The oscillatory motions are said to be *out of phase* with each other. A measure of how much out of phase the two motions are is the difference, in time, between the occurrence of the maximum values of the two graphs. This is approximately 0.005 s. The amplitude of the oscillations is 0.6 radians and the period is again 0.06 s.

The function $\theta = 0.6[2\pi(t - 0.005)/0.06]$ can be used as a simple model for the motion of the forewings. Again it is only an approximation; the graph of this function again does not fit the data exactly (see Fig. 1.3).

The simple functional forms presented here provide simple models but Figs 1.2 and 1.3 show that some improvement is needed for a good model. We could improve the representation in each case by choosing a combination of sine and cosine functions and finding a Fourier series—but this is beyond the scope of this book.

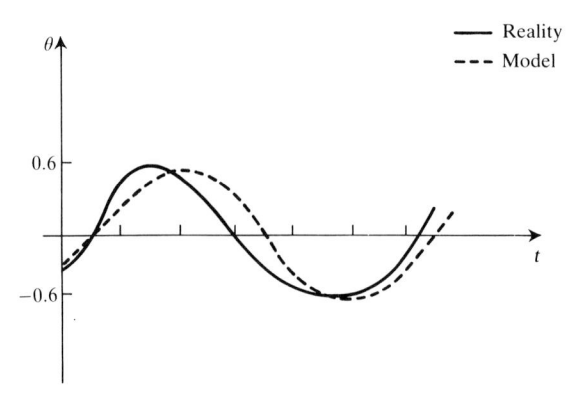

Fig. 1.3

• Exercise

During the process of breathing, the pressure in the lungs is a periodic function. This function is shown in Fig. 1.4 (where p_0 is the maximum pressure). When we breath in (inspiration) the pressure drops and when we breath out (expiration) the pressure increases. Use a sine function as a simple model to represent this pressure. Sketch this sine function on top of the graph of the pressure so as to show the approximate nature of the model.

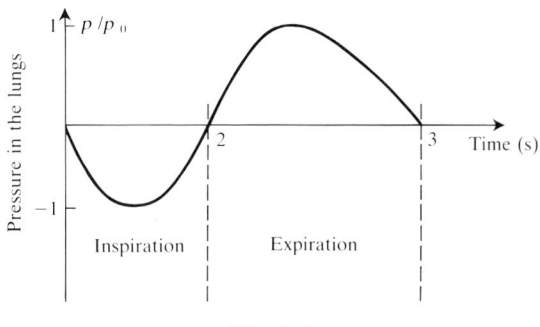

Fig. 1.4

• Further reading

De Sapio, R. (1978). *Calculus for the life sciences*, W. H. Freeman & Co., San Francisco.

Functions

1.2. Party representation and election results

- **Polynomial functions; graph fitting of data**

The relationship between the number of votes cast in an election and the number of seats obtained in the legislative body by the participating parties is a subject of interest to politicians and political scientists. The number of seats a particular party receives in parliament is obviously a function of the votes cast, but can this function be represented in mathematical terms?

Figure 1.5 shows the results of elections to the Irish Dail between 1923 and 1973. Find a linear model between the proportion of seats won and the proportion of votes cast for the Fianna Fáil Party.

- **Solution**

Perhaps somewhat surprisingly the relationship between votes and seats is reasonably linear in the range represented by the data. Figure 1.6 shows a graph of the data of Fig. 1.5.

The equation of the straight line is

$$y = 1.2x - 0.05$$

where y is the proportion of seats won and x is the proportion of votes cast.

Functions

DAIL ELECTIONS (1923–1973)

Year	Party	% of votes	Seats won	Year	Party	% of votes	Seats won
1923	C na G	39.2	63	1948	F.F.	41.9	67
	Sinn Fein	27.6	44		F.G.	19.8	31
	Labour	11.6	14		Labour	8.7	14
	Others	21.6	32		Others	29.6	34
1927 (June)	C na G	27.4	46	1951	F.F.	46.3	68
	F.F.	26.1	44		F.G.	25.7	40
	Labour	12.6	22		Labour	11.5	16
	Others	33.9	40		Others	16.6	22
1927 (Sept)	C na G	38.4	61	1954	F.F.	43.4	65
	F.F.	34.9	57		F.G.	32.0	50
	Labour	8.9	13		Labour	12.1	18
	Others	12.4	14(+3)*		Others	12.5	13
1932	F.F.	44.6	72	1957	F.F.	48.3	78
	C na G	35.3	56		F.G.	26.6	40
	Labour	7.7	7		Labour	9.1	11
	Others	12.4	14(+3)*		Others	16.0	17
1933	F.F.	49.7	76	1961	F.F.	43.8	70
	C na G	30.5	48		F.G.	32.0	47
	Labour	5.7	8		Labour	11.6	15
	Other	14.1	20		Others	12.6	11
1937	F.F.	45.3	68	1965	F.F.	47.7	72
	F.G.	34.8	48		F.G.	34.1	47
	Labour	10.0	13		Labour	15.4	21
	Others	9.9	8		Others	2.8	3
1938	F.F.	52.0	72(+4)	1969	F.F.	45.7	74
	F.G.	33.3	43(+2)		F.G.	34.1	50
	Labour	10.0	9		Labour	17.0	18
	Others	4.7	7		Others	3.2	1
1943	F.F.	41.9	66	1973	F.F.	46.2	68
	F.G.	23.1	32		F.G.	35.1	54
	Labour	15.7	17		Labour	13.7	19
	Others	19.3	22		Others	5.0	2
1944	F.F.	48.9	73(+2)				
	F.G.	20.6	29(+1)				
	Labour	8.9	8				
	Others	21.6	24		* Seats in brackets are unopposed.		

Fig. 1.5

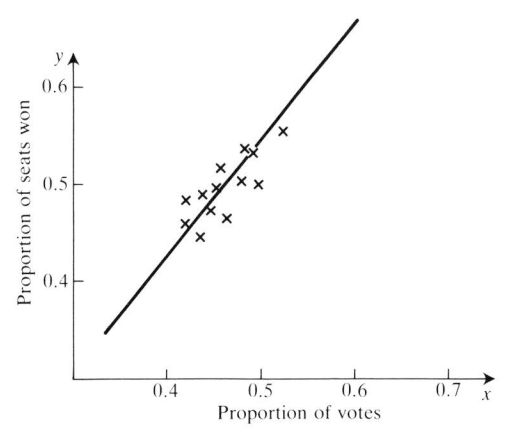

Fig. 1.6

Functions

- **Exercise**

1. Table 1.1 shows the election results for the British parliament (at Westminster). British political scientists interested in a relationship between the proportion of votes cast and the number of seats in a two-party system (i.e. Conservative/Labour parties) use a cube law

Table 1.1. British general elections

Year	Conservative (per cent)		Liberal (per cent)		Labour (per cent)		Others (per cent)	
	votes	seats	votes	seats	votes	seats	votes	seats
1918	35	54	23	23.5	15	10	27	12.5
1922	39	56	29	18.5	28.5	23	2.5	2.5
1923	38	42	29.5	26	30.5	31	2	1
1924	47	67	18	6.5	33	24.5	2	2
1929	38	42	23.5	10	37	47	1.5	1
1931	55	76	11	12	30	8.5	4	3.5
1935	54	70	6.5	3	37.5	25	2	2
1945	40	33	9	2	48	62	3	3
1950	43	47.8	9	1.5	46	50.4	2	0.3
1951	48	50.5	2.5	1	48.5	47	1	0.5
1955	49.8	55	2.7	0.9	46.3	44	1.2	0.2
1959	49.4	58	5.9	0.9	43.8	41	0.9	0.1
1964	43	43	11	1.4	44	50.3	5.3	0.1
1966	42	40	8.5	2	48	58	1.7	0.1
1970	46.4	52.5	7.5	1	43	45.5	3.2	1

$$\frac{y}{1-y} = \left(\frac{x}{1-x}\right)^3 \tag{1.3}$$

where x is the proportion of votes cast and y is the proportion of seats won. Draw a graph of this law and include the data from Table 1.1. How good a model is this cube law?

1.3. Growth rates

• **Exponential laws, logarithmic laws**

Under favourable conditions, in a controlled biological experiment, cell division may take place at very nearly equal intervals of time. We shall investigate the case of reproduction by 'binary fission', as illustrated in Fig. 1.7. (In the analysis, it will be helpful to use a *step function* defined by

$$[x] = \text{the greatest integer not exceeding } x.$$

For example,

$$[5.84] = 5, \quad [\pi] = 3, [\sqrt{2}] = 1, \quad [4] = 4,$$
$$[-3.7] = -4, \quad [-1] = -1.)$$

To take an example, a bacterium called *Escherichia coli*, resident in human intestines, reproduces by binary fission at the rate of one division approximately every 17 minutes, at body temperature; slower if the temperature is lower. For ease of calculation, let us take the period between division to be 20 minutes. Find a formula relating the population of cells P to time t.

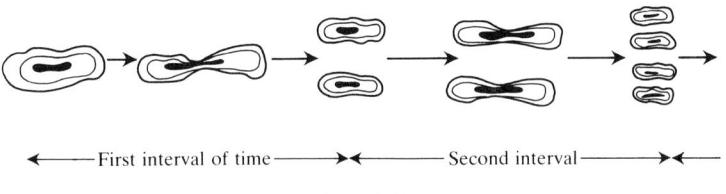

←———First interval of time———→←———Second interval———→←—

Fig. 1.7

• **Solution**

Assuming synchronous cell division throughout, we can construct Table 1.2 for the population. From Table 1.2 we have the population P after time t minutes given by the formula

$$P = 2^{[t/20]}.$$

Now if we started our experiment with a colony of bacteria instead of a single specimen, we should not expect the first divisions to be exactly synchronous. In fact it would be more realistic to assume an *average* rate of one division every 20 minutes thereafter. In such a case, starting with, say, 1000 bacteria, we have an approximate doubling of population over any interval of 20 minutes, with an associated formula

$$P = 1000 \times 2^{t/20}.$$

A more convenient expression for this relation for some purposes is

$$\log P = \log 1000 + \frac{t}{20} \log 2.$$

Table 1.2

Time interval t (min)	Population
0–20	1
20–40	2
40–60	4
60–80	8

Functions

The graph of log P against t is linear, with gradient $(1/20)\log 2$. Figure 1.8 shows graphs of P and log P plotted against time. Of course each of these curves is continuous and this would not model accurately the real population which would be discrete.

$\log P = 1000 \times 2^{t/20}$

$\log P = \log 1000 + \dfrac{t}{20} \log 2$

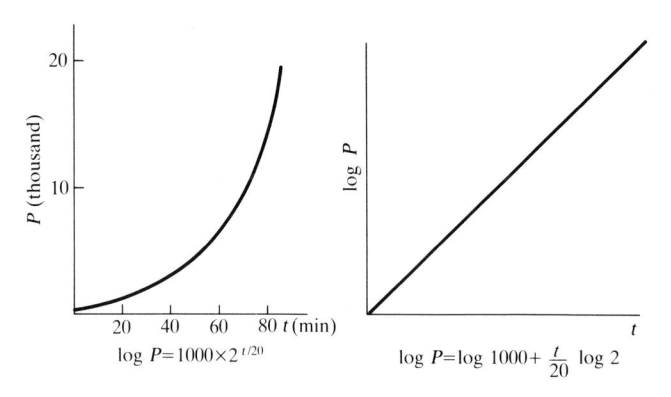

Fig. 1.8

• Exercises

1. A hill-farmer is building up a flock of sheep and decides to keep all his ewe lambs (and sell off all the ram lambs not needed for breeding purposes). A ewe is able to give birth at the age of 12 months and the average number of lambs per ewe is 1.2 each year, of which half are ewes and half rams. Suppose there are no losses due to weather or other causes. The flock initially consists of 100 ewes, and the first lambs are born a year later.

   (i) Find the number of *ewes*, P, in the flock after t years.

   (ii) Express log P in terms of t.

   (iii) After how many years does P exceed 600?

2. The following table gives the population of Europe at intervals of 50 years.

Year	1700	1750	1800	1850	1900	1950
Pop. (millions)	106	117	156	214	303	392

Let t be the number of years after 1700 and P be the number of millions of population. Plot log P against t and estimate (by eye) the line of best fit. Hence estimate the population in the year 2000.

1.4. Apparent magnitudes of stars

• **Logarithmic laws**

Astronomers of 2000 years ago, of whom Hipparcus (150 BC) was one of the greatest, catalogued the brightness of the stars, according to their apparent brightness, i.e. how bright the star appeared to the naked eye. The brightest stars were classified as first-magnitude stars, the next brightest stars as second magnitude, and so on.

By the mid-nineteenth century a 'visual photometer' was invented which could measure the brightness of stars very accurately.

If we take the magnitude of the *brightest* star of the first magnitude as classified by Hipparcus and call it *magnitude 1* and take the magnitude of the *brightest* star of the second magnitude and call it *magnitude 2*, then a star of apparent brightness halfway between the stars of magnitude 1 and 2 is assigned the magnitude 1.5.

Table 1.3 shows the brightness ratio, i.e. the ratio of the actual brightness of two stars and the corresponding magnitude differences. Find a relation between $m_2 - m_1$ and S_1/S_2.

Table 1.3

Magnitude differences ($m_2 - m_1$)	1	2.5	5	10	15	20
Brightness ratio (S_1/S_2)	2.512	10	100	10^4	10^6	10^8

• **Solution**

The relation between $(m_2 - m_1)$ and S_1/S_2 obeys a simple logarithmic law. We have

$$m_2 - m_1 = 2.5 \log_{10}\left(\frac{S_1}{S_2}\right)$$

$$\frac{S_1}{S_2} = 10^{0.4(m_2 - m_1)}.$$

Functions

1.5. Shortest paths across rectangular cities

- **Binomial coefficients**

Fig. 1.9

Figure 1.9 is an aerial view of Manhattan, New York. As the picture shows the roads of Manhattan (called *avenues* in one direction and *streets* in the perpendicular direction) are laid out in the form of a rectangular grid. This feature of road layout is common to many American cities.

Work out how many different routes there are between two specified junctions.

- **Solution**

Figure 1.10 shows a simple schematic diagram which can be used to define the problem.

Consider a city with roads running north/south and east/west, such that the distance between each junction (i) going east is x metres and (ii) going south is y metres.

Suppose that we want to go from A to B by the shortest distance and we are constrained to go only along the roads. The shortest distance is $(4x + 3y)$ metres but there are several possible routes. Two such routes are shown in Fig. 1.11.

Functions

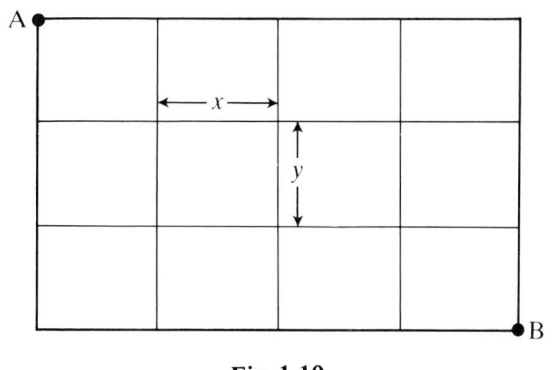

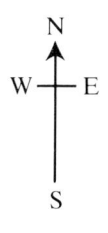

Fig. 1.10

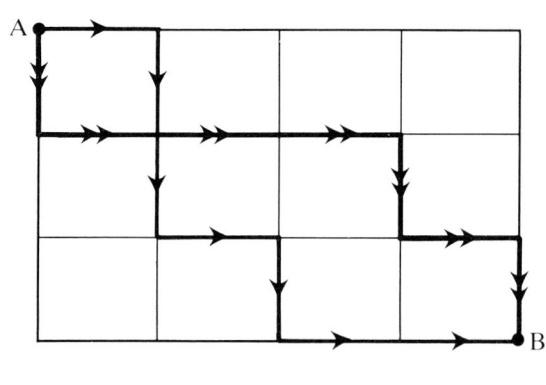

Fig. 1.11

The problem is: how many routes are there? We can write out each possible path in the form of the following table.

| E | S | S | E | S | E | E |

This is the route in Fig. 1.11 marked by single arrow heads.

Another route is shown by the double arrow heads in Fig. 1.11 and can be written as

| S | E | E | E | S | E | S | .

The total number of possible routes is the same as the number of ways of choosing 3 Ss and 4 Es to fill the 7 boxes in the tables. The total number of ways of doing this is 35 and this number arises in the expansion of $(x + y)^7$: 35 is the coefficient of $x^3 y^4$. So the real problem of finding the number of ways of travelling between two junctions gives an example of the binomial coefficients ^nC_r (here we have $^7C_3 = 35$).

This problem illustrates the role played by certain numbers in mathematics. The problem can be localized if a local town or city is built on a grid system

• Further reading

Open University Press (1979). O.U. Course MS283, Block I, Unit 2.

Functions

- **Exercise**

How many routes are there between
  (i) A and C
  (ii) C and B
in the grid shown in Fig. 1.12? Write these as binomial coefficients.

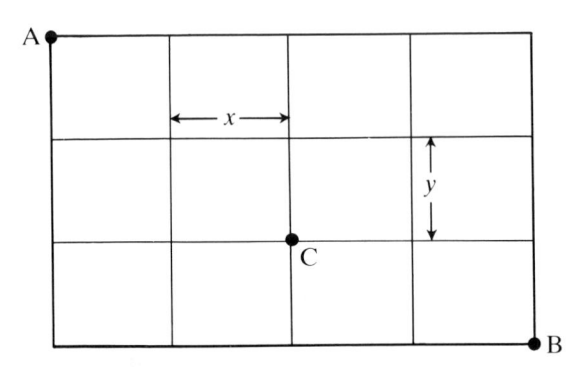

Fig. 1.12

Functions

1.6. Gravitational force

• **Binomial theorem**

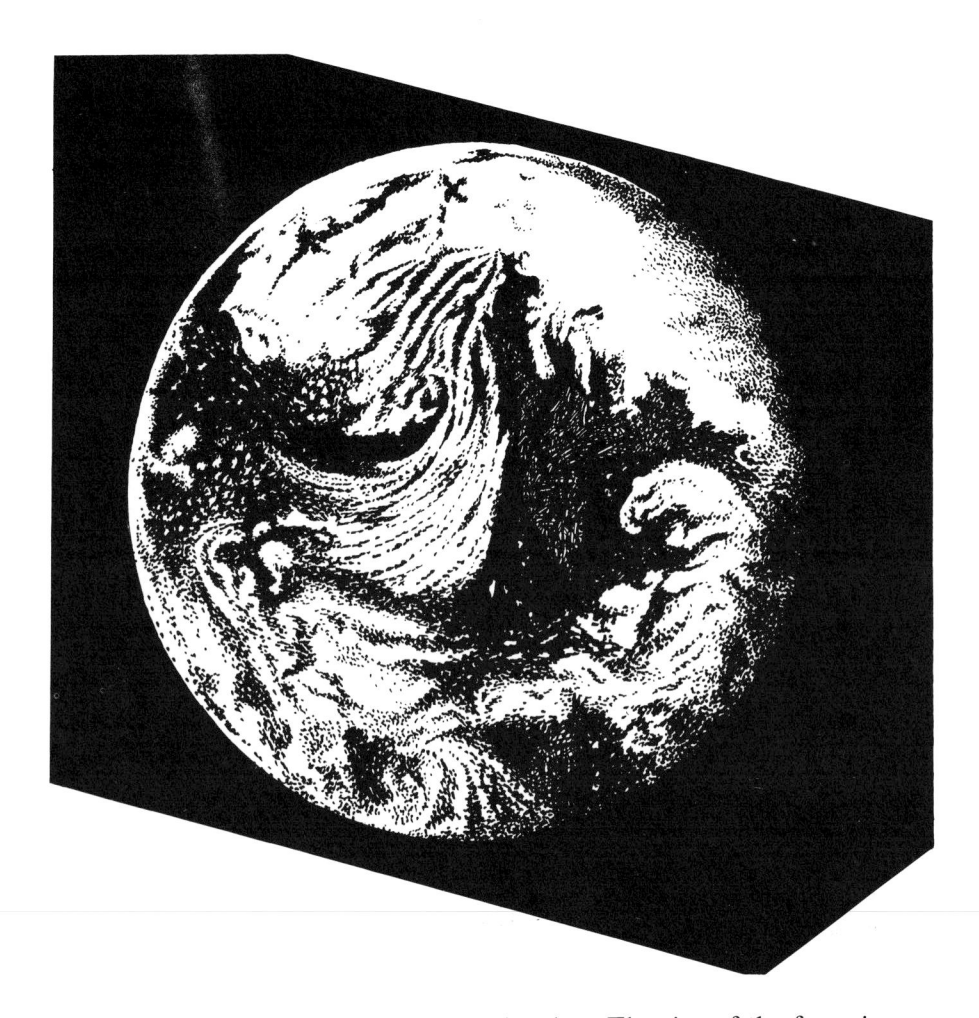

Any two bodies exert a force of attraction on each other. The size of the force is

$$\frac{Gm_1 m_2}{r^2}$$

where m_1 and m_2 are the masses of the two bodies, r is the distance between their centres of mass, and G is a constant of value 6.668×10^{-8} dyn cm² g⁻². The force is negligible unless one of the bodies is very large. In particular, if a body is near to the earth's surface the force is called the weight of the body, and is taken as a constant. If a body is a small distance δ from the earth's surface, show that the gravitational force may be taken as a constant to within 2 per cent of its true value for $\delta \leqslant 40$ miles.

$$\text{Data: } M = 5.977 \times 10^{27}\,\text{g}$$
$$a = 6.378 \times 10^3\,\text{km} \ (\simeq 4000 \text{ miles}).$$

• **Solution**

Suppose that we model the earth as a sphere of radius a and of mass M. Consider a body of mass m and a distance δ from the earth's surface (see Fig. 1.13).

Functions

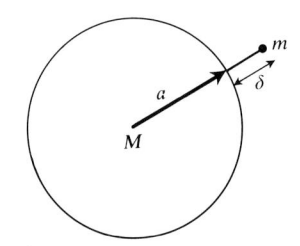

Fig. 1.13

The force F on the body is given by

$$F = \frac{GMm}{(a+\delta)^2}$$

which can be written as

$$F = \frac{GMm}{a^2\left(1+\dfrac{\delta}{a}\right)^2}.$$

For a body on the earth's surface, $\delta = 0$ and

$$F = \left(\frac{GM}{a^2}\right)m.$$

Now we are used to writing $F = mg$ so we now have a theoretical expression for g, the acceleration due to gravity. We have

$$g = \frac{MG}{a^2}$$

and putting the values for G, M, and a we find

$$g = 980.2 \text{ cm s}^{-2}.$$

If the body is not on the earth's surface, then the weight of the body is no longer constant. Instead we can write,

$$\text{Weight} = m\frac{GM}{a^2\left(1+\dfrac{\delta}{a}\right)^2} = \frac{mg}{\left(1+\dfrac{\delta}{a}\right)^2}$$

$$= mg\left(1+\frac{\delta}{a}\right)^{-2}.$$

Table 1.4 shows how the weight of a body decreases as the body is moved away from the earth's surface. Thus we have to go some way from the earth's surface before the weight of a body decreases by much; for instance when 40 miles from the earth the weight is reduced by 1/50 or 2 per cent.

Table 1.4

δ		Weight (mg)
(km)	(miles)	
6.378	4	0.998
63.78	40	0.98
637.8	400	0.826

Functions

In fact a body can go 216 miles up before its weight is reduced by 10 per cent. The formula for the weight of the body can be expanded using the binomial theorem as follows

$$\text{Weight} = mg\left(1 + \frac{\delta}{a}\right)^{-2}$$

$$= mg\left(1 - \frac{2\delta}{a} + 3\left(\frac{\delta}{a}\right)^2 - 4\left(\frac{\delta}{a}\right)^3 + \ldots\right).$$

Thus the first- and second-order approximations to the weight are, respectively,

$$W_1 = mg\left(1 - 2\frac{\delta}{a}\right)$$

and

$$W_2 = mg\left(1 - 2\frac{\delta}{a} + 3\left(\frac{\delta}{a}\right)^2\right).$$

The graph in Fig. 1.14 shows these two approximations.

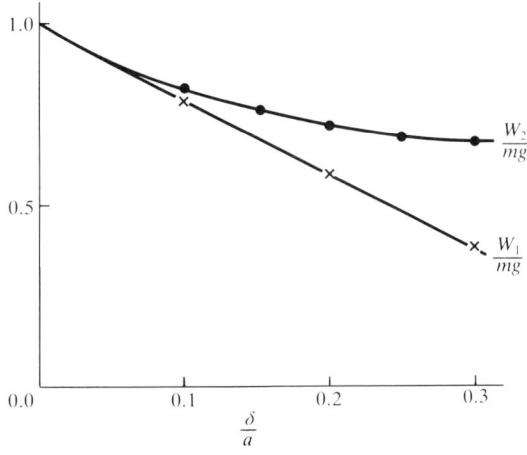

Fig. 1.14

• Exercise

What is the third-order approximation to the weight of a body of mass m? For what distance, δ, does the third-order approximation differ from the second-order approximation by 10 per cent?

Functions

1.7. Investment and borrowing

- **Arithmetic series**

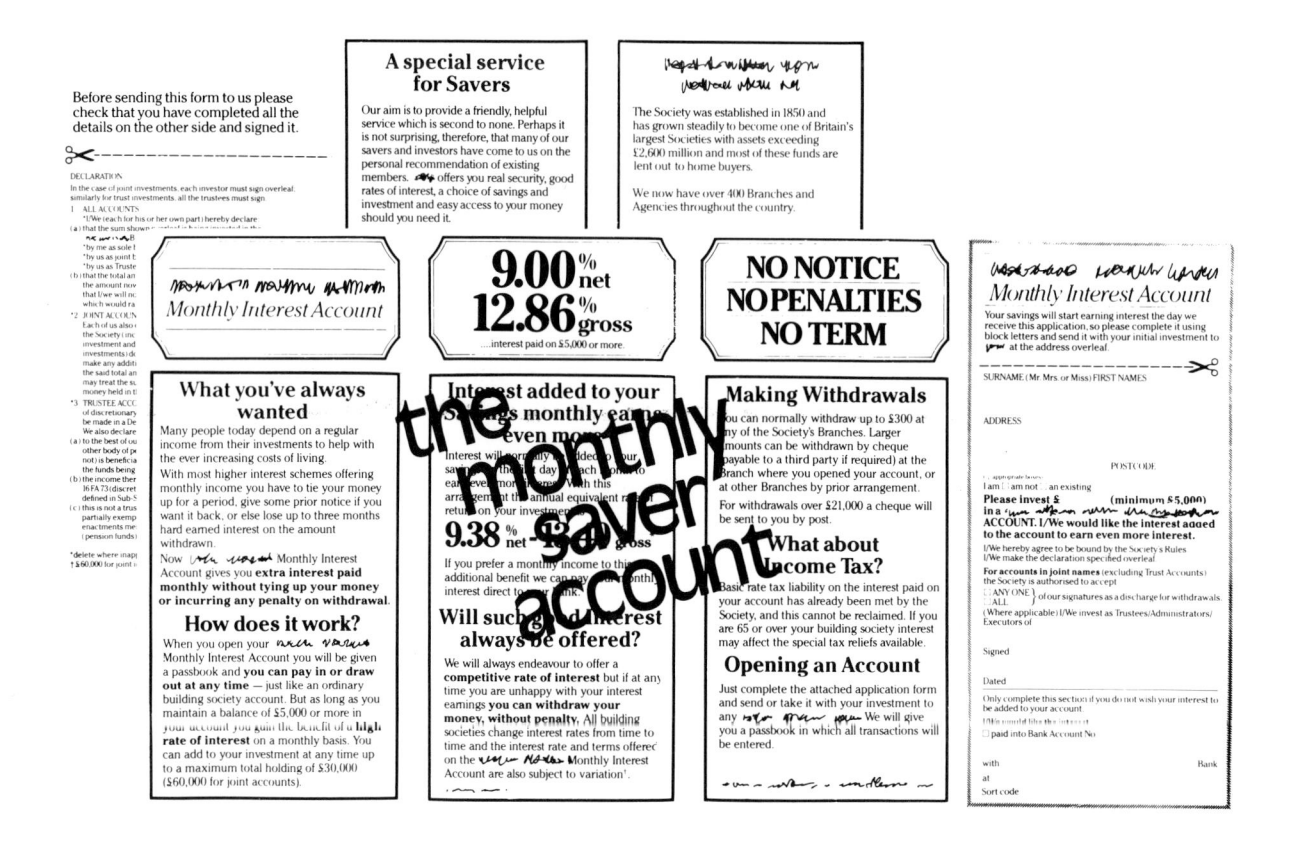

Most people save at sometime in their life, using a variety of means: saving accounts, building societies, unit trusts, endowment insurance, stocks and shares, and so on. Many of us also borrow money: loans, hire purchase, credit agreements, mortgages. Saving, or investment, is the same transaction, basically, as borrowing but seen the other way round.

If you invest money in any way, you are really lending your money and for this you are paid, usually at a fixed rate of interest. If you borrow money, you pay to do so, also usually at a fixed rate of interest.

The usual pattern of saving is for some initial amount, the principal, to be invested, followed by regular equal payments. Let us assume, for instance, that the payments are made annually. Using the following notation

I = initial principal in £s

p = regular (annual) payment in £s

r = rate of interest (per cent)

A_n = accumulated amount after n years in £s,

find a formula for A_n in terms of I, r, p, and n.

Functions

- **Solution**

Letting $R = r/100$, for simplicity, we have

$$\text{Year } 0: A_0 = I$$
$$\text{Year } 1: A_1 = I(1 + R) + p$$
$$\text{Year } 2: A_2 = [I(1 + R) + p](1 + r) + p \quad I(1 + R)^2 + p(1 + R) + p$$
$$\text{Year } 3: A_3 = [I(1 + R)^2 + p(1 + R) + p](1 + R) + p$$
$$= I(1 + R)^3 + p(1 + R)^2 + p(1 + R) + p.$$

Extending the pattern, we get

$$\text{Year } n: A_n = I(1 + R)^n + p[1 + (1 + R) + (1 + R)^2 + \ldots + (1 + R)^{n-1}].$$

If we can find a simple way of calculating

$$[1 + (1 + R) + (1 + R)^2 + \ldots + (1 + R)^{n-1}],$$

then we will have a simple formula for A_n, the accumulated amount after n years.

Let $S_n = 1 + (1 + R) + (1 + R)^2 + \ldots + (1 + R)^{n-1}$ (there are n terms in the sum).

Multiplying throughout by $(1 + R)$ gives

$$(1 + R)S_n = (1 + R) + (1 + R)^2 + \ldots + (1 + R)^{n-1} + (1 + R)^n.$$

Subtracting

$$RS_n = (1 + R)^n - 1.$$

Therefore,

$$S_n = \frac{(1 + R)^n - 1}{R}.$$

Therefore,

$$A_n = I(1 + R)^n + pS_n$$

$$= I(1 + R)^n + \frac{p}{R}[(1 + R)^n - 1]$$

$$= \left(I + \frac{p}{R}\right)(1 + R)^n - \frac{p}{R}.$$

- **Exercises**

1. Suppose a family puts down £100 and pays £100 per annum into a 5 per cent investment trust. Show that after 20 years they should receive £3572 (to the nearest pound)

2. A man deposits £150 annually to accumulate at 10 per cent compound interest. How much money will he have to his credit immediately after making the tenth deposit?

3. A man saves £200 each year and invests it at the end of the year at 8 per cent compound interest. How much will the combined savings and interest amount to at the end of 15 years?

4. In the case of house purchase, I is a negative amount, and p represents the annual repayment. Suppose a family wishes to buy a £20 000 house over 25 years. The building society

Functions

demands a 20 per cent down payment and interest is charged at 14 per cent per annum on the outstanding loan. What would the monthly repayment be?

5. *A bouncing ball*: A rubber ball dropped from a height h rebounds to a height $2h/5$. Find the total distance covered before the ball comes to rest.

1.8. Scree slopes

• **Area of triangle, trigonometric formulae**

The erosion of cliff-faces may result in the formation of scree at the base of the cliff. By making measurements on the remaining cliff-face and scree it may be possible to calculate how far the cliff has retreated under erosion from its original position. Figure 1.15 shows the theoretical shape of a cliff face at two different times.

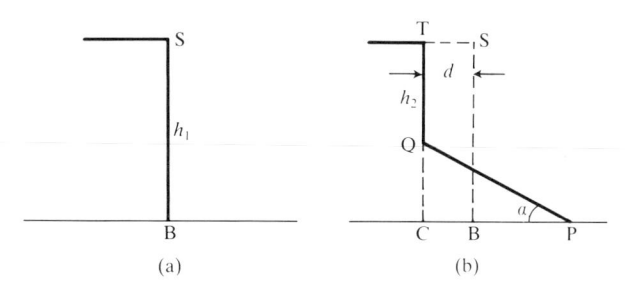

Fig. 1.15

We assume that the cliff-face is vertical and that initially it is of the same height, h_1, throughout its length. The ground at the foot is level.

When scree forms it makes a slope of uniform gradient, inclined at an angle α to the horizontal. Figure 1.15(a) shows the initial cliff profile and Fig. 1.15(b) the profile after erosion. Find a formula for the angle α in terms of $h_1, h_2,$ and d.

• **Solution**

The areas SBCT and CPQ shown in Fig. 1.15(b) must be equal.

Area of SBCT $= h_1 d$.

Area of CPQ $= \frac{1}{2} \text{CP} \cdot \text{CQ}$

Now CQ $= h_1 - h_2$ and CP $= \text{CQ}/\tan \alpha = (h_1 - h_2)/\tan \alpha$

Functions

So area of CPQ = $(h_1 - h_2)^2/(2 \tan \alpha)$.

Equating these areas and dividing by h_1 we have

$$d = \frac{(h_1 - h_2)^2}{2h_1 \tan \alpha}.$$

• Exercises

1. Find α if $d = 2$, $h_1 = 50$, and $h_2 = 30$.

2. Obviously, many refinements of the model are possible. An easy refinement is illustrated in Fig. 1.16.

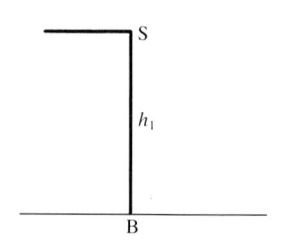

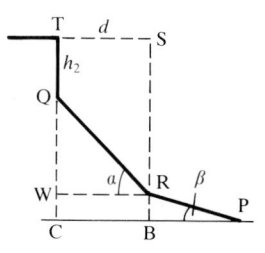

Fig. 1.16

Again, letting the length of retreat be d, initial height h_1, and final height h_2, find a quadratic equation for d in terms of h_1, h_2, α, and β.

We could take the gradient of the scree slope to be variable, with profile a circular arc, tangential to the level ground and the cliff-face. Try this if you want a harder problem.

2. Calculus

The problems in this chapter use calculus as part of the solution. §§ 2.1–2.3 are maximum/minimum type problems, and §§ 2,4–2.8 involve simple first-order ordinary differential equations. § 9 contains a selection of models in the field of population dynamics; these involve various types of ordinary differential equation.

2.1. The shape of a tin can

Many types of food are sold in tin cans, but do manufacturers use the most economical design for their cans? Most cans are cylindrical, although certain foods, such as pressed meats, are canned in a cuboid. What shape should such a cylindrical can be? For instance, the dimensions of a 500 g can of tomatoes are diameter 115 mm and height 130 mm, whereas a 250 g can has dimensions, diameter 67 mm and height 130 mm. Why this change in shape? Is it using a lot more metal than necessary? Design a can that requires the minimum amount of material for a fixed volume.

- **Solution**

The total surface area of a cylindrical can is

$$A = \tfrac{1}{2}\pi D^2 + \pi D h$$

where D is the diameter of the base and h is the height. The volume is

$$V = \tfrac{1}{4}\pi D^2 h.$$

The expression for surface area contains two variables, D and h. Thus in order to use a calculus method to discover the dimensions giving minimum area, we must first eliminate one variable. From

$$V = \tfrac{1}{4}\pi D^2 h,$$

we have

$$h = 4V/\pi D^2$$

so that

$$A = \tfrac{1}{2}\pi D^2 + \pi D \frac{4V}{\pi D^2}$$

$$= \tfrac{1}{2}\pi D^2 + \frac{4V}{D}.$$

To find the minimum area we now differentiate this with respect to D and set the answer equal to zero.

$$\frac{dA}{dD} = \pi D - \frac{4V}{D^2}$$

$$= 0 \text{ for maximum or minimum,}$$

Calculus

$$\text{i.e. } D^3 = 4V/\pi \quad \text{and so} \quad D = \left(\frac{4V}{\pi}\right)^{1/3}.$$

To check that this is indeed a minimum we use the second derivative test. Thus

$$\frac{d^2 A}{dD^2} = \pi + \frac{8V}{D^3}$$

which is positive when $D = \left(\dfrac{4V}{\pi}\right)^{1/3}$,

and so this value of D does indeed give minimum surface area. The corresponding value of the height is

$$h = \frac{4V}{\pi D^2} = \frac{\pi D^3}{D\pi^3} = D.$$

Minimum material is used when the height and diameter of a cylindrical are equal, and this minimum surface area is given by

$$\tfrac{1}{2}\pi D^2 + \pi D^2 = \tfrac{3}{2}\pi D^2.$$

- **Exercise**

Find the minimum surface area of a cuboid with a square base and a given volume.

2.2. Fleet size for car leasing company

The leasing section of the Spode Car Hire and Leasing Company leases cars to large corporations on a yearly basis. It charges £1000 per car per year, but for contracts with a fleet size of more than 10 cars the fee per car is discounted by 1 per cent for each car in excess of 10. The company wishes to know how many cars it must lease to a single corporation to produce maximum *income*. It also wishes to know about *profits*. Assuming that each car depereciates by £500 in the year, find the number of cars leased to a single corporation which will produce the maximum profit for the company.

• **Solution**

If I denotes the total *income* for a contract with a fleet size x, then

$$I = 1000x, \qquad \text{if } 0 \leq x \leq 10$$
$$= (1000 - 1000(x - 10)/100)x, \quad \text{if } x > 10$$

since for $x > 10$ the rental for every car is reduced by 1 per cent of £1000 for each car in excess of 10 in the contract. The problem is to determine the fleet size x which maximizes I. In the range $0 \leq x \leq 10$, I is a linearly increasing function. Thus the maximum value of I is obtained at the endpoint $x = 10$, where $I = 10\,000$. For $x > 10$ we have

$$I = (1000 - 1000(x - 10)/100)x$$
$$= (1000 - 10(x - 10))x$$
$$= 1100x - 10x^2.$$

The maximum value of I can be found by setting $dI/dx = 0$. We have

Calculus

$$\frac{dI}{dx} = 1100 - 20x$$

$$= 0 \text{ for maximum or minimum.}$$

Thus $x = 55$.

This is indeed a maximum since $d^2I/dx^2 = -20 < 0$. In fact, $x = 55$ gives an income of £30 250, so this value is the maximum for $x \geq 0$ also. In theory the company could lease up to 110 cars to a large corporation and still have an income from the corporation. This can be seen by considering the total income function,

$$I = (1100 - 10x)x, \quad \text{for} \quad x > 10.$$

Thus, $I > 0$ for $x < 110$. The situation is illustrated in Fig. 2.1.

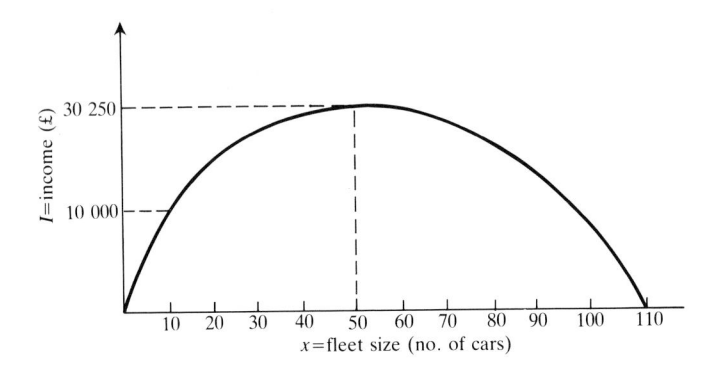

Fig. 2.1

The company is more interested in the fleet size that will produce maximum *profit* rather than maximum *income*, however. The depreciation on each car in the fleet is £500 each year, and so the profit function P is given by

$$P = 1000x - 500x, \quad \text{if } 0 \leq x \leq 10$$
$$= 1100x - 10x^2 - 500x, \quad \text{if } x > 10.$$

In the range $0 \leq x \leq 10$, P is a linearly increasing function and hence has its maximum of £5000 at the endpoint $x = 10$.

For $x > 10$ we have
$$P = 600x - 10x^2.$$

Then
$$\frac{dP}{dx} = 600 - 20x$$

$$= 0 \text{ for maximum or minimum.}$$

Thus $x = 30$ is a stationary point, and

$$\frac{d^2P}{dx^2} = -20$$

confirms that it is a maximum.

For $x > 10$, a fleet size of 30 cars produces a maximum profit P of £9000. This value is greater than the £5000 above and so is a maximum for all x. The situation is illustrated in Fig. 2.2.

Calculus

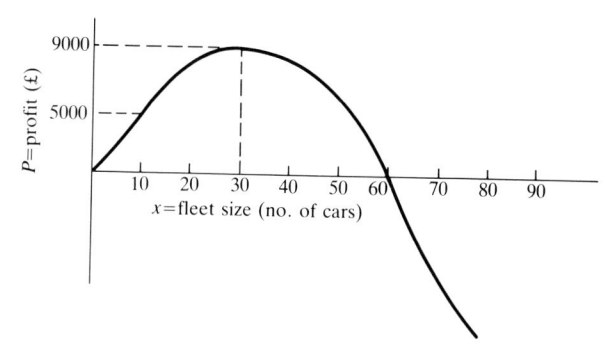

Fig. 2.2

It should be noted that the fleet size for maximum income, $x = 55$, produces only a small profit, £2750, and fleet sizes in excess of 60 produce a loss. The company is well advised either

1. to ensure the maximum fleet size is around 30, or
2. to change the discount rate.

• Exercise

Repeat the problem with a least charge per car of £1500 instead of £1000. Does this affect the fleet size for maximum income and maximum profit?

Calculus

2.3. Stock control

Most shops and warehouses hold a considerable amount of stock in hand, and know the annual demand for it with a good degree of accuracy. The sales will clearly be subject to random fluctuations about a trend, and the pattern of stock level with time will follow a pattern of the form shown in Fig. 2.3.

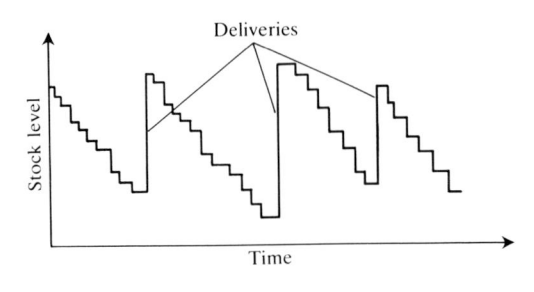

Fig. 2.3

Especially in times of economic depression, the owner of the shop or warehouse will want to ensure that he is not keeping valuable stock on the shelves for longer than is necessary. Thus he will want to use an optimal stockholding and purchasing policy for his products.

- **Solution**

We approach this problem retaining some generality and insert the data at the end.

The relationship displayed in Fig. 2.3 is clearly too complicated to model exactly, so we will first make the simplifying assumption that the owner has a demand spread evenly over the year and so orders an amount Q regularly every time period T. We also assume that it is important for him not to run out of stock completely at any time, so that he always keeps an amount s in stock. This gives the situation illustrated in Fig. 2.4, where Q is known as the lot size, T as the re-order period, and s as the safety stock.

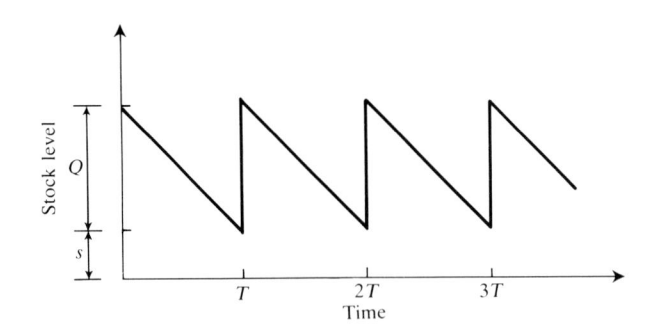

Fig. 2.4

Looking at this simplified graph, we see that we have assumed that the stock level is repeatedly dropping from $s + Q$ to s in a linear manner. Thus the average stock held throughout the year is $s + \frac{1}{2}Q$. Also, with a re-order period of T years there will be $1/T$ deliveries per year.

We now assume that the total cost of stockholding each year depends on *three* factors:

1. the cost of buying the item;
2. the administrative cost of ordering a delivery;
3. the cost of keeping the item in stock (things like rent, rates, storemen's wages).

Introducing the notation,

N = annual demand in units per year

C_1 = order cost per order

C_2 = purchase cost per item

a = stockholding cost (as fraction of price) per item,

enables us to write the total annual cost as

$$C = \frac{1}{T}C_1 \quad + \quad NC_2 \quad + \quad \left(s + \frac{Q}{2}\right)aC_2.$$
$$\text{Order} \qquad\qquad \text{Purchase} \qquad\qquad \text{Stockholding}$$

We also note that T and Q are related via $(1/T)Q = N$ so that

$$C = \frac{NC_1}{Q} \quad + \quad NC_2 \quad + \quad \tfrac{1}{2}aC_2(2s + Q).$$
$$\text{Order} \qquad\qquad \text{Purchase} \qquad\qquad \text{Stockholding}$$

Sketching each of these costs separately against the lot size Q gives Fig. 2.5, from which it is clear that the total cost C should have a minimum for a certain value of Q.

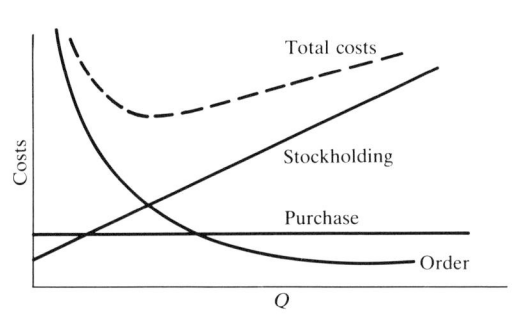

Fig. 2.5

To find this minimum cost we differentiate C with respect to Q and set this equal to zero.

$$\frac{dC}{dQ} = -\frac{NC_1}{Q^2} + 0 + \tfrac{1}{2}aC_2$$

$$= 0 \text{ when } Q = (2NC_1/aC_2)^{\frac{1}{2}}.$$

This value of Q, at which the total annual cost will be a minimum, is known as the economic batch quantity (EBQ) or economic order quantity. The corresponding value of T is

$$T = \frac{Q}{N} = (2C_1/aNC_2)^{\frac{1}{2}}.$$

These EBQ formulae form the basis for nearly all stockholding and purchasing policies.

• Exercises

1. Suppose that a car distributor expects to cell 10 000 oil filters this year, and that they cost him £2 each. He knows that there is a cost of £5 incurred each time he places an order, and an annual cost of holding the filter in stock of about 18 per cent of its price. Show that this annual cost will be minimized if he orders 521 filters every 19 days.

Calculus

2. For the situation above, in practice it may well be necessary to order filters fortnightly or monthly. Show that this means ordering about 834 or 417 filters, respectively. Show that choosing a monthly re-order period involves an extra cost of about £20, whereas a fortnightly one only involves about £5 excess.

- **Further reading**

Similar models are discussed in the following books.

Burghes, D. N. and Wood, A. D. (1980). *Mathematical models in the social management and Life Sciences.* Ellis Horwood.

Burghes, D. N., Huntley, I. D., and MacDonald, J. J. (1982). *Applying mathematics: a course in mathematical modelling.* Ellis Horwood.

2.4. Sales response to advertising

Most firms that advertise their products do so with the advice of an advertising agency. The agency has a great deal of practical experience, and uses this to predict how the firm's sales will respond to an advertising campaign. We investigate this here.

In the absence of any sort of promotion, sales figures tend to drop—purchasers are perhaps lured away by the attractive advertisements of similar products, or may just wish to try something new. Examples of this are shown in Figs 2.6 and 2.7, where the second graph, especially, shows a seasonal effect as well. The data have been displayed on a log–linear graph, which has the effect of making the steady decrease in sales very nearly linear.

The advertising campaign will be designed to turn this steady decrease in sales into a steady increase. This increase is unlikely to continue for long, however, and saturation will result; this is shown in Fig. 2.8, where sales tend to stabilize after an initial period of growth.

Formulate a model for the sales response to advertising which reproduces these features.

- **Solution**

The approach given here is in the form of a differential equation. We will ignore the seasonal effect visible in Figs 2.6 and 2.7, and assume that the linear decrease shown is accurate. Introducing the notation

$$s = \text{sales rate}$$

$$t = \text{time},$$

we can now write

$$\ln s = a - bt$$

Calculus

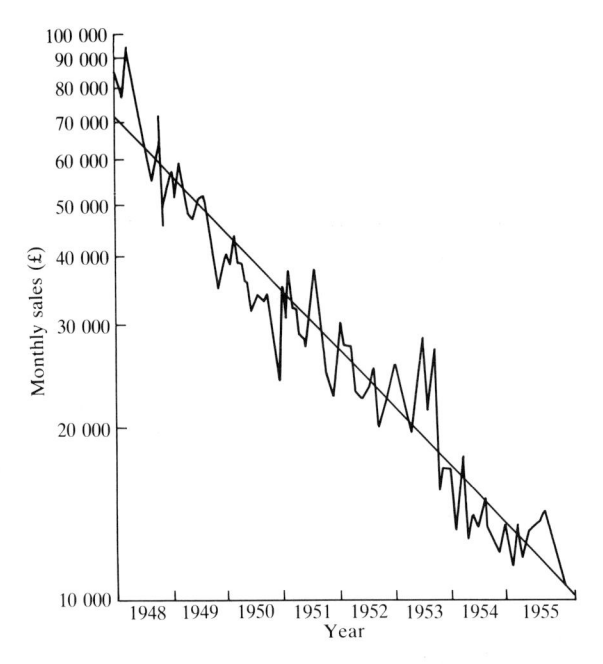

Fig. 2.6

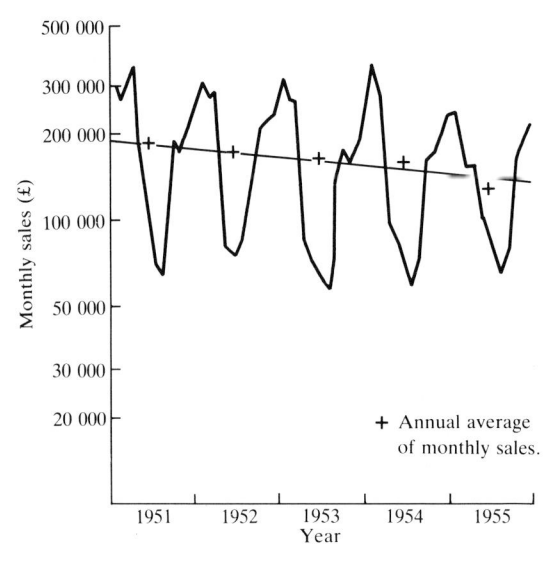

Fig. 2.7

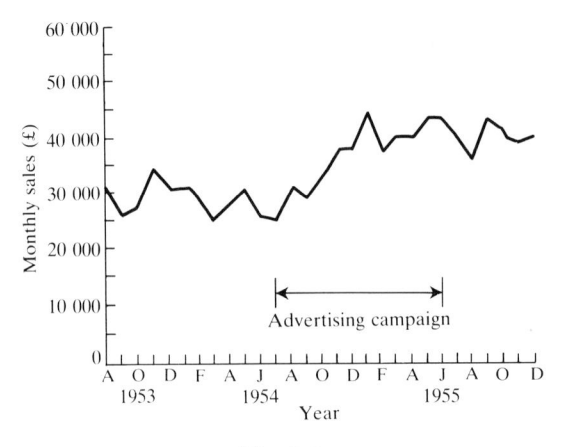

+ Annual average of monthly sales.

Fig. 2.8

Advertising campaign

34

where the constants a and b can be obtained from the linear graph. Thus we immediately have

$$\frac{ds}{dt} = -bs$$

when there is no advertising.

We now want to include the effects of advertising. We know that an increased advertising rate A leads to an increased sales rate S, and so, for simplicity, we choose

$$\frac{ds}{dt} \propto A.$$

We also want to account for the amount of product already sold: clearly if the sales rate s is approaching the saturation level M (the practical limit of sales), it will require a lot of advertising to change the sales very much. For simplicity we choose

$$\frac{ds}{dt} \propto (M-s).$$

Combining these two assumptions with the original model leads to the differential equation

$$\frac{ds}{dt} = cA\frac{(M-s)}{M} - bs$$

where c is a constant and the M in the denominator is purely for later convenience. Rearranging this equation gives

$$\frac{ds}{dt} + \left(\frac{cA}{M} + b\right)s = cA,$$

a linear first-order differential equation whose solution will clearly depend on the advertising rate A.

To complete the problem we must choose a particular form for $A(t)$. As an example, we consider the simple case where A is constant for a campaign of length T and then zero. Thus

$$A(t) = \begin{cases} \bar{A}, & 0 < t < T \\ 0, & t > T. \end{cases}$$

We also assume that the sales rate is non-zero at the start of the campaign, so that

$$s(0) = s_0.$$

Then our equation becomes

$$\frac{ds}{dt} + fs = c\bar{A} \quad \text{for} \quad 0 < t < T,$$

where the constant f is given by

$$f = \frac{c\bar{A}}{M} + b.$$

This may be solved either by separating the variables or by using the integrating factor e^{ft}. The latter method gives

$$e^{ft}s = \int e^{ft} c\bar{A}\, dt$$
$$= \frac{c\bar{A}}{f} e^{ft} + D$$

where D is a constant of integration. Hence

$$s(t) = \frac{c\bar{A}}{f} + De^{-ft},$$

Calculus

and use of the initial condition $s(0) = s_0$ gives

$$s_0 = \frac{c\bar{A}}{f} + D.$$

Therefore,

$$s(t) = \frac{c\bar{A}}{f} + \left(s_0 - \frac{c\bar{A}}{f}\right)e^{-ft} \quad \text{for} \quad 0 < t < T.$$

Now for $t > T$, the advertising rate $A = 0$. Thus the differential equation is

$$\frac{ds}{dt} + bs = 0,$$

which has solution

$$s(t) = \beta e^{-bt} \quad \text{for} \quad t > T$$

and some constant β. These two solutions must be equal when $t = T$, and this enables us to find β

$$s(T) = \frac{c\bar{A}}{f} + \left(s_0 - \frac{c\bar{A}}{f}\right)e^{-fT} = \beta e^{-bT}.$$

We finally obtain the sale rate as

$$s(t) = s_0 e^{-ft} + \frac{cA}{f}(1 - e^{-ft}), \quad \text{if} \quad 0 < t < T$$

$$= \left[s_0 e^{-fT} + \frac{cA}{f}(1 - e^{-fT})\right] e^{-b(t-T)}, \quad \text{if} \quad t > T$$

where

$$f = \frac{c\bar{A}}{M} + b.$$

This is sketched in Fig. 2.9.

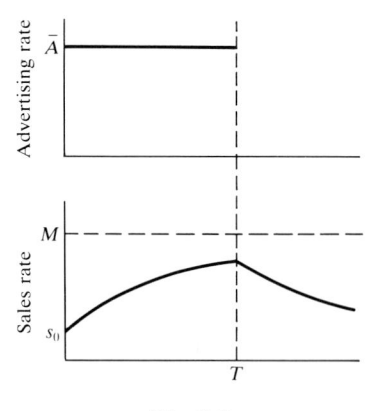

Fig. 2.9

It can be seen that the sales rate increases while the advertising campaign is in operation, but decreases as soon as the advertising ceases at time T.

- **Exercise**

Analyse the sales response when $A(t) = \alpha t$, where α is a positive constant and we are only concerned with the case $b = 0$.

2.5. Radioactive decay

Radioactive decay is the process whereby an unstable atom disintegrates to form a new element (which may also be radioactive) whilst emitting radiation. This radiation can be in the form of moving subatomic particles of electromagnetic radiation. For example, carbon 14, a radioactive element, will decay to the stable element carbon 12. Both these substances (carbon 12 and its isotope carbon 14) are present in living things, and hence also in historical objects made of once-living material. Carbon-dating is a method of approximating the date of construction of an object made from natural materials, by coupling the theory of radioactive decay with the ability to measure the concentration of the isotope carbon 14 in the object.

If the rate of change of the number of undecayed atoms is proportional to this number of atoms, find a general formula for N as a function of time and interpret any constants in your formula.

• **Solution**

Our assumption (actually postulated by Rutherford and others earlier this century) can be written symbolically as

$$\frac{dN}{dt} \propto N$$

where $N(t)$ is the number of undecayed atoms present at time t. Thus

$$\frac{dN}{dt} = -kN$$

where k is a positive constant, and so

$$\int_{N_0}^{N} \frac{dN}{N} = -\int_{0}^{t} k\,dt$$

where N_0 is the number of atoms present at a fixed time $t = 0$ and $N_0 \neq 0$.

$$[\ln N]_{N_0}^{N} = -[kt]_0^t$$

$$\ln N - \ln N_0 = -kt + 0$$

$$\frac{N}{N_0} = e^{-kt}.$$

Thus our model is

$$N = N_0 e^{-kt} \tag{2.1}$$

The constant k is called the decay constant for the substance. To complete the model for a given radioactive material we need to find this constant; this obviously has to be done experimentally. However there is a slicker but indirect way of doing this. Suppose that, at a particular time t_1, there is a quantity N of radioactive atoms present, and at a time t_2 there is a quantity $\frac{1}{2}N$ present; then

$$\tfrac{1}{2}N = N e^{-k(t_2 - t_1)}$$

i.e.

$$e^{-k(t_2 - t_1)} = 2.$$

Taking logs to base e we get

$$t_2 - t_1 = \frac{\ln 2}{k}. \tag{2.2}$$

Calculus

Here it can be seen that the time for half of a given quantity to decay is *constant*, no matter what this amount is. We call this constant time, $\tau = t_2 - t_1$, the half-life of the substance. Hence, if we can measure the time for any quantity of a substance to decay to half of the amount then we can evaluate k from the relation:

$$k = \frac{\ln 2}{\tau} \quad \text{(see eqn (2.2))}.$$

It is illuminating to look at the graph of a general exponential function of the form $y = a e^{-kt}$ (Fig. 2.10). From it one can easily appreciate the important constant feature of an exponential function, namely the half-life.

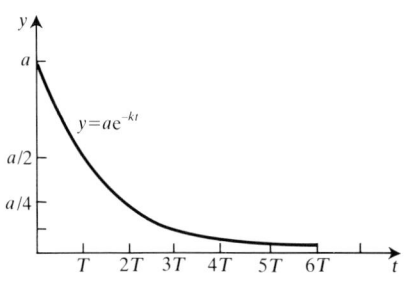

Fig. 2.10

Let us look at the calculations involved in a simple example. The half-life of barium 140 is about 13 days. If initially there is l_0 grams of barium 140 present in a sample,

1. how much will remain after two hours?
2. how long will 2/3 of the barium 140, present initially, take to decay?

First we must find k, the decay constant. From above we have

$$k = \frac{\ln 2}{\tau} = \frac{\ln 2}{13}.$$

(Note, all time measurements must now be in *days*.)

1. l_0 g was initially present, so, comparing with eqn (2.1), we have $N_0 = l_0$, $k = \ln(2)/13$, and two hours later $t = 1/12$ (of a day). Thus when $t = 1/12$

$$N = l_0 e^{\left(-\frac{\ln 2}{13} \cdot \frac{1}{12}\right)} = 0.996 \, l_0$$

2. Time is implicit in eqn (2.1), so we take logarithms to the base e, resulting in

$$\ln N(t) = \ln N_0 - kt$$

$$\ln \frac{N(t)}{N_0} = -kt.$$

We have $N_0 = l$, $N(t) = \tfrac{1}{3} l_0$ (i.e. the amount *left* after an unknown T days), $k = \ln(2)/13$. Thus

$$T = \frac{-2}{(\ln(2)/13)} \ln(\tfrac{1}{3} l_0 / l_0).$$

$$= + \frac{13}{\ln 2} \ln 3$$

$$= 20.6 \text{ days}.$$

Finally, it should be mentioned that this theory relating to radioactive decay has been widely verified experimentally, and we will be justified in the next section in using this model to study the decay of carbon 14.

• Exercises

1. Xenon 133 has a half-life of about 5 days. Find the decay constant for xenon 133. If a sample contains 5 g of radioactive xenon 133, how much would have decayed after 10 days? How long would it take for 10 per cent of a given quantity of xenon 133 to decay?

2. The decay constant for strontium 90 is approximately 2.77×10^{-2} (day^{-1}). Calculate the half-life of strontium 90, and hence find the time for 1 kg to decay until only 0.25 kg remain.

3. Uranium 238 has a half-life of 4.5×10^9 years. Calculate the *initial* decay rate of 1 kg of uranium 238. Find the amount of uranium 238 in the 1 kg sample when the decay rate has reduced to 10^{-10} kg per year.

2.6. Carbon dating

This technique for dating objects was developed by an American, W. F. Libby, in the 1940s, and is widely used today to detect the authenticity of an historical object. The method is not sufficiently accurate to give exact dates for objects; its use often lies in deciding whether an article is of modern origin or from antiquity.

Carbon 14 is formed in the atmosphere by the coalition of free neutrons and nitrogen atoms. The neutrons are formed by cosmic rays (high-energy electromagnetic radiation) bombarding the atmosphere. The premise upon which the technique of carbon-dating relies is that the bombardment has been constant throughout time—a point about which scientists will argue. The carbon 14 is absorbed by plants and animals during their lives, and an equilibrium state is reached whereby the rate of absorption equals the rate of decay. Only when the plant or animal dies does the level of carbon 14 begin to decline, due to the lack of further absorption. Briefly, the method is to measure the rates of decay of the carbon 14 present in the historical object, and in a piece of living material from which the object is made, and to use the theory from §2.5 to estimate the date of manufacture. (*Note*, we cannot date the manufacture of an article made from already antique material (as many forgeries are!).)

- **Solution**

Suppose we wish to find the approximate date of manufacture of an oak chair. We need to find the rates of decay of the carbon 14 present in the chair, and in living oakwood. We shall call these rates R_c and R_w, respectively, where t is the time, in years, from the date of manufacture to the present day. It is the value of t which we are trying to find; t is the time taken for the decay rate of the carbon 14 to reduce from R_w to R_c. We know from §2.5 (see eqn (2.1)) that

$$\dot{N}(t) = R(t) = -kN(t) \quad (k > 0).$$

Now from the equation $N(t) = N(0)e^{-kt}$ we have

$$\frac{-R(t)}{k} = -\frac{R(0)}{k}e^{-kt}$$

$$R(t) = R(0)e^{-kt}. \tag{2.3}$$

Here, as in the previous example concerning barium 140 (see §2.5), time is implicit in the equation, and an identical analysis yields

$$t = \frac{-1}{k}\ln\left(\frac{R(t)}{R(0)}\right) = -\frac{1}{k}\ln\left(\frac{R_c}{R_w}\right) = \frac{1}{k}\ln\left(\frac{R_w}{R_c}\right).$$

We do not as yet know the value of k, but this can easily be found from τ, the half-life of carbon 14, which is known to be approximately 5568 years. Using this, we get

$$k = \frac{\ln 2}{5568}.$$

Hence, we have

$$t = \frac{5568}{\ln 2}\ln\left(\frac{R_w}{R_c}\right).$$

We can now easily find the age t, of the chair, from the measured values of R_w and R_c. As an example of the simple calculation required we shall look at some real figures by considering the 'round table' in Winchester Castle. This table is supposed to be the original round table of King Arthur, who was alive in the fifth century AD. Let us see if the table really is 1500 years old.

From measurements taken, R_w and R_c for this table are approximately 6.68 and 6.08 respectively. Thus, we have

$$t = \frac{5568}{\ln 2} \cdot \ln\left(\frac{6.68}{6.08}\right) \approx 756 \text{ years.}$$

Obviously, the table was not King Arthur's!

• **Exercises**

1. If the half-life of carbon 14 is 5568 years, and the rate of decay of carbon 14 in living wood is known to be about 6.68, estimate the ages of antique wooden objects exhibiting the following rates of decay of the carbon 14 present in them

(i) 6.45; (ii) 5.80; (iii) 2.12.

All the decay rates are per minute per gram of the sample.

2. Consider again the 'round table' of the example given above of carbon-dating. Suppose that in the measurement of the two decay rates, i.e. for the living wood (6.68) and for the wood of which the table is made (6.08), there is the possibility of a 5 per cent error (either way). Is it conceivable that the table *could* be King Arthur's?

Calculus

2.7. Art forgeries

This section is about an application of the theory of radioactive decay. It is not set as a problem, as the applications are in other sections, but should provide interesting reading for the student.

Many art forgeries so closely resemble the originals that very high-powered scientific techniques are needed to verify that a painting is indeed a forgery. In fact, many paintings are now scrutinized as a matter of course, the forging process often being a fine art itself. Even with these techniques, many eminent art dealers and scientists often dispute the authenticity of a particular painting. To cite an example of how good art forgery can be, there is the case of H. A. Van Meegaran. This Dutch painter was imprisoned in 1945 for collaborating with the Nazis in selling Goering a seventeenth-century painting, 'Woman taken in adultery', by Vermeer. He then claimed, from his prison cell, to have painted the picture himself; along with some other famous pieces of art including 'Disciples of Emmaus', by Vermeer. After much investigation and debate by a distinguished panel of scientists and art historians, the paintings were in fact pronounced forgeries, and Van Meegaran was sentenced to one-year imprisonment for his crime. However, many did not believe the evidence, and, to illustrate the doubt which hung over the methods used, the painting 'Disciples at Emmaus' was recently sold for $170 000!

More recently a method has been developed which allows fairly conclusive proof as to whether a painting is authentic. This method has since shown the self-confessed forgeries of Van Meegaran to be, in fact, very good copies of the originals. The method involves the presence of a radioactive substance called white lead (lead 210) in the pigments from which paint is made. The ores from which white lead is extracted also contain other radioactive substances such as uranium and in particular radium 226. Lead 210 is the product when radium 226 decays. During manufacture of the pigment most of the radium and its descendants are removed, but not completely. The small amount of radium 226, which survives the manufacturing process, then begins to decay to lead 210 and, since it decays far more slowly than the lead 210, an equilibrium state is reached between the two elements. It is this equilibrium state which we wish to analyse.

To be able to understand the following one will need to be familiar with the process of radioactive decay (see §§ 2.5 and 2.6).

Two important pieces of information are needed before we begin. These are the half-lives of radium 226 and lead 210, which are 1600 years and 22 years, respectively.

Let $P(t)$ be the amount of lead 210 per gram of ordinary lead at time t and let $P(t_0)$ be the amount of lead 210 per gram of ordinary lead at the time of manufacture t_0. Let $R(t)$ be the number of disintegrations per minute of radium 226 per gram of ordinary lead, at time t. Let k be the decay constant for lead 210, $k > 0$.

At time t, we have a rate of decay for the lead 210 of $-kp(t)$. However, the radium 226 is also decaying, *producing* lead 210, at a rate $R(t)$. Thus we have a net rate of decay of the lead 210 in the pigment of $-kp(t) + R(t)$. We can write this, as

$$\frac{dp}{dt} = -kp + R(t). \tag{2.4}$$

From now on we shall spend some time in solving eqn (2.4) to find $p(t)$ explicitly. This will be our model for the concentration of lead 210 in a paint pigment at any time t after the time of manufacture. One must bear in mind that, like most models, only includes certain variables, and the reader must question whether significant ones have been ignored in composing this model.

Equation (2.4) is a first-order differential equation in which the variables are not separable, but it may be solved by the integrating factor method since the equation is linear. If we have an equation of the form

$$a_1(x)\frac{dy}{dx} + a_0(x)y = h(x),$$

then the solution of the equation is more easily found by the multiplication of each term by the integrating factor

$$\exp\left(\int \frac{a_0}{a_1} dx\right).$$

For eqn (2.4), the integrating factor is therefore $e^{\int k\,dt} = e^{kt}$. Hence,

$$e^{kt} \cdot \frac{dp}{dt} = -e^{kt}kp + e^{kt}R$$

and this can be written as

$$\frac{d}{dt}(e^{kt}p) = e^{kt}R.$$

Integrating with respect to t gives

$$e^{kt}p = \int e^{kt}R\,dt + C$$

where C is constant. Thus

$$p = e^{-kt}\int e^{kt}R\,dt + e^{-kt}C. \tag{2.5}$$

This is as far as we can go, as we do not know the nature of $R(t)$ and hence cannot integrate $e^{kt}R(t)$. Remember, though, that the half-life of radium 226 is 1600 years and, if we are investigating paintings only a few hundred years old, the radium will have decayed only a little and thus, to all intents and purposes, $R(t)$ will have remained roughly constant. Bear in mind that we are not trying to *date* paintings, but just trying to distinguish between originals and forgeries. If $R(t)$ is constant, R_0 say, we have

$$p = e^{-kt}R_0\frac{1}{k}e^{kt} + Ce^{-kt}$$

and so

$$p(t) = \frac{R_0}{k} + Ce^{-kt}.$$

We also know that at $t = t_0$, $p(t) = p(t_0)$. Thus,

$$p(t_0) = \frac{R_0}{k} + Ce^{-kt_0}$$

and so

$$C = e^{kt_0}\left(p(t_0) - \frac{R_0}{k}\right).$$

Our final model now becomes

$$p(t) = \frac{R_0}{k} + \left(p(t_0) - \frac{R_0}{k}\right)e^{-k(t-t_0)}. \tag{2.6}$$

From eqn (2.6) we wish to find $t - t_0$. We know, or can measure, all the other quantities present in the equation except $p(t_0)$. Before manufacture, the original quantity of lead 210 was in equilibrium with the larger quantity of radium 226 in the ore. We shall call the rate at which this larger quantity of radium decays B. We know two things about B. First we are able to measure this rate now, and readings show that its value varies over the range 0–200 disintegrations

Calculus

per gram per minute. Second, the original quantities of lead 210 and radium 226 were in equilibrium, and thus $B = kp(t_0)$. Our strategy is to *suppose* that the painting under investigation *is* authentic, and to insert the corresponding value of $t - t_0$ into eqn (2.6). We shall then rearrange eqn (2.6) to find $kp(t_0)$ (and hence B) and investigate the value of B resulting from this.

For example, if we choose $t - t_0 = 400$, i.e. for a 400-year-old painting, we have

$$p(t) = \frac{R_0}{k} + \left(p(t_0) - \frac{R_0}{k}\right)e^{-400k}$$

$$kp(t) = R_0 + (kp(t_0) - R_0)e^{-400k}$$

$$kp(t)e^{400k} = R_0 e^{400k} + kp(t_0) - R_0$$

$$kp(t_0) = kp(t)e^{400k} - R_0(e^{400k} - 1).$$

Hence,

$$B = kp(t_0) = kp(t)e^{400k} - R_0(e^{400k} - 1). \tag{2.7}$$

We now insert the known values of k (3.15×10^{-2}), R_0, and $p(t)$. If B is outside the range 0–200 we can assume that the 400-year-old painting is in fact a forgery.

As an example, let us return to the painting 'Disciples at Emmaus' by Van Meegaran/Vermeer. Measurements show that for this painting $kp(t) = 8.5$ and $R_0 = 0.8$. We shall assume that the painting is genuinely from the seventeenth century, and thus, $t - t_0 \simeq 300$. Equation (2.7) now becomes

$$B = 8.5\, e^{300 \times 3.15 \times 10^{-2}} = 0.8(e^{300 \times 3.15 \times 10^{-2}} - 1)$$

$$= 97\,853.$$

This value of B is far too high (remember, it ought to be in the range 0–200!), and so we assume that the 'Disciples at Emmaus' is indeed a forgery.

• Exercises

1. The values of $kp(t)$ and R_0 were measured from a collection of five supposedly sixteenth-century paintings. These data, for the five paintings A–E, follow. Which do you consider forgeries?

	A	B	C	D	E
$kp(t)$	4.3	2.1×10^{-5}	12.6	2.1257	6.97×10^{-5}
R_0	0.7	1.2×10^{-5}	3.2	2.1258	4.16×10^{-5}

2. Equation (2.7) can be rearranged into the form

$$B = kp(t_0) = (kp(t) - R_0)e^{k(t-t_0)} + R_0.$$

If we choose $t = 0$ to be the current time, then $t - t_0$ becomes $-t_0$. Hence we have

$$B = (kp(0) - R_0)e^{-kt_0} + R_0.$$

The value of t_0 is negative so that, for example, for a 400-year-old painting $t_0 = -400$.

Suppose now that you are investigating a suspect 200-year-old painting, i.e. $t_0 = -200$. Sketch a graph of $p(0)$ against R_0 supposing that B takes its upper limit of 200. On the same graph sketch $p(0)$ against R_0 again, but this time for $B = 0$. Indicate a region corresponding to the allowable values of B, i.e. $0 \leq B \leq 200$.

2.8. Drug absorption

For a drug to be used safely two important pieces of information are needed. One is the dosage level, which will vary according to the effect required, and will obviously be bounded by a maximum safe dose. The other, with which we will be concerned, is the time interval between doses. Once the required dose has been ascertained by research and experimentation, the pharmacologist's job is to instruct the user what the frequency of the dose must be. To solve this problem we must consider how the body deals with a drug, once in contact with tissue, i.e.

1. the absorption of the drug;
2. its excretion via various organs of the body.

During the ensuing discussion it will be obvious that many assumptions are being made. Obviously every drug and every recipient has its quirks, but we will simplify the real situation in order to obtain a model involving the mathematics available.

ABSORPTION

By absorption we mean the process of the drug being distributed through the tissues of the body. Normally this process is accomplished very quickly. Some reasons for this follow.

1. The total blood capacity of an average 70-kg man is around 6 litres while the heart circulates around 5 litres of blood per minute. Thus, in a little over a minute, the entire volume of blood in the body makes one circuit.
2. Drugs are normally required to be absorbed into the blood as quickly as possible, so as to facilitate fast action; this often involves intravenous injection or oral intake.

Against these arguments we have the fact that the blood is only about 7 per cent of the total body liquid, and so complete absorption of a drug thus takes much longer than the minute intimated above. However, even considering this, drug absorption is still relatively fast, especially when compared with the time for the excretion of the drug.

DRUG CLEARANCE

Once absorbed, the process of removing the drug from the body tissues begins. This process, referred to as clearance, is accomplished by a variety of organs in the body, foremost being the kidneys. Removal by the kidneys, or renal clearance, releases the drug via the bladder into the urine, and it is from urine tests that the rate of renal clearance of a drug can be determined. Some drugs cannot be excreted by renal clearance at all, some only by this means. However, we shall use results obtained from research carried out on renal excretion to extrapolate to clearance in general. The assumption which we shall make is that *the renal clearance rate is proportional to the quantity of the drug in the body tissues*. We shall assume that this is true for drug clearance in general. We have bounds set on the renal clearance rate. Since the amount of blood plasma flowing past the kidneys is about 0.65 litres per minute, we have a range of 0–0.65 litres per minute for the clearance rate.

When a drug is administered, its toxicity will immediately decline, due to clearance and, if a constant level is to be maintained, frequent further doses will be required. As an example of the quite rapid rate at which the concentration of a drug in the body decreases with time Fig. 2.11 shows results for penicillin for various initial doses. The problem is to administer the drug in such a way that the concentration remains within the required therapeutic range.

Find an equation that describes the concentration of a drug at time t. Use it to suggest methods of drug administration.

Calculus

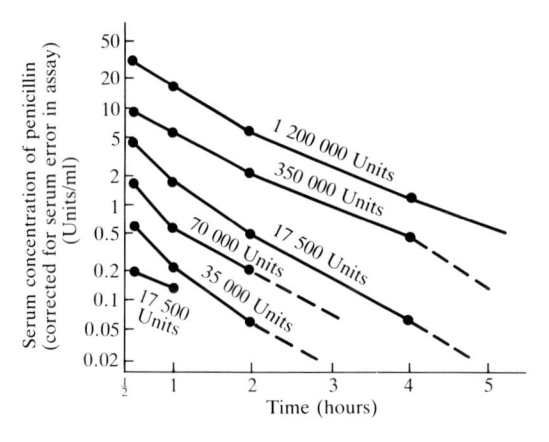

Fig. 2.11

• **Solution**

Let $y(t)$ be the quantity of a drug in the bloodstream and body tissues at a time t. Then the discussion above leads us to

$$\frac{dy}{dt} \propto y,$$

i.e.

$$\frac{dy}{dt} = -ky \qquad (2.8)$$

where k is positive constant.

By an analysis with which you should now be familiar (see §§ 2.5–2.7), eqn (2.8) yields the general solution

$$y = y_0 e^{-kt}. \qquad (2.9)$$

Here y_0 is the initial dose of the drug given at time $t = 0$. Suppose, now, that at a time $t = T$ a second identical dose of the drug is given, and then again at times $2T$, $3T$, etc. Also we shall denote the instants just before doses are given as $T-$, $2T-$, $3T-$, etc., and the instants just after doses are given as $T+$, $2T+$, $3T+$, etc. The sequence of quantities of the drug in the bloodstream at times 0, $T-$, $T+$, $2T-$, $2T+$, etc. are

$$y(0) = y_0$$
$$y(T-) = y_0 e^{-kT} \quad \text{(from eqn (2.9))}$$
$$y(T+) = y_0 + y_0 e^{-kT} \quad \text{(dose of } y_0 \text{ given)}$$
$$= y_0(1 + e^{-kT})$$
$$y(2T-) = y_0(1 + e^{-kT}) e^{-kT}.$$

(Note, the above equation corresponds to eqn (2.9) with y_0 replaced by the new initial dose of $y_0(1 + e^{-kT})$, i.e. the quantity of the drug present at time $T+$.)

$$y(2T+) = y_0 + y_0(1 + e^{-kT}) e^{-kT} \quad \text{(another dose of } y_0 \text{ given)}$$
$$= y_0(1 + e^{-kT} + e^{-2kT}).$$

Continuation of this procedure yields the result that at time $t = nT+$ we have

$$y(nT+) = y_0(1 + e^{-kT} + \ldots e^{-nkT}). \qquad (2.10)$$

The bracketed portion of this equation is a geometric progression with a first term of 1 and a common ratio of e^{-kt}. The sum of this progression of $n + 1$ terms is

$$\frac{1 - e^{-(n+1)kT}}{1 - e^{-kT}}.$$

Hence we have

$$y(nT+) = \frac{y_0(1 - e^{-(n+1)kT})}{1 - e^{-kT}}. \tag{2.11}$$

If the dose is administered many times then, as n becomes large,

$$e^{-(n+1)kT} \to 0.$$

Thus

$$y(nT+) \to \frac{y_0}{1 - e^{-kT}}$$

where $y_0/(1 - e^{-kT})$ is a constant and represents the maximum quantity of the drug that can build up in the body from a succession of doses of quantity y_0 at intervals of time T. We shall call this amount the saturation level, denoted by y_s. For a given saturation level of the drug in the bloodstream we can choose either the dosage or the period between the doses, and calculate the other. However, we have not yet discussed how to find the constant k, analogous to the decay constant in radioactive decay.

From eqn (2.9) we have $y/y_0 = e^{-kt}$, so that

$$\ln(y/y_0) = -kt.$$

Let $t = 0$, when $y = y_0$, and $t = \tau$ when $y = \tfrac{1}{2}y_0$; then at $t = \tau$,

$$\ln(\tfrac{1}{2}y_0/y_0) = -k\tau.$$

This gives

$$\tau = -\frac{\ln(\tfrac{1}{2})}{k} \tag{2.12}$$

or

$$k = \frac{\ln 2}{\tau}. \tag{2.13}$$

τ is the time required for the quantity of the drug to reduce from y_0 to $y_0/2$, when no further dose is administered and, as can be seen from (2.12) above, is constant. τ is called the halftime of elimination. Compare τ with its equivalent in radioactive decay theory where it is the half-life of an element.

Now, to find k (from eqn (2.13)) we need to find a value for τ, which is usually done experimentally.

In practice, the method of administering a fixed dose at set intervals of time leads to a very slow build-up to the saturation level, y_s. A solution to this problem is to start with one large dose, and then to proceed with smaller doses. An obvious choice for this large dose is y_s, then the next dose y_d brings the level again to be y_s where

$$y_s = y(T+) = y_s e^{-kt} + y_d$$

and so $y_d = y_s(1 - e^{-kt})$. Thus $y_d = y_0$ (since we know that $y_s = y_0/(1 - e^{-kt})$).

Thus a saturation level of y_s can be achieved by an initial dose of y_s and then successive doses of y_0. This method of administering a drug, and the previous one of giving successive equal doses of y_0, are illustrated in Fig. 2.12. Clearly the latter method is an improvement.

Calculus

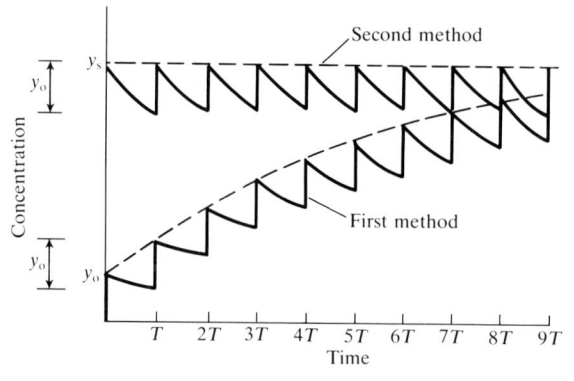

Fig. 2.12

As an example suppose that a particular drug is to be administered to a patient over a period, with a required constant level of 500 mg. It is known that the half-time of elimination for the drug is 75 minutes. How much of the drug needs to be administered every 5 hours in order to maintain a level of 500 mg?

We must calculate y_0, that is, the dose to be administered at intervals after the first 500 mg. Referring to the symbolism used in the analysis, we know τ, y_s, and T. First we calculate a value for k.

From eqn (2.13)
$$k = \frac{\ln 2}{\tau}$$
$$-\frac{\ln 2}{75}.$$

Now from eqn (2.11) we know that
$$y_s \leq \frac{y_0}{(1 - e^{-kt})}.$$

Hence
$$y_0 \simeq y_s(1 - e^{-kt}).$$

With $k = \ln(2)/75$, $y_s = 500$, and $t = 300$ minutes, we get
$$y_0 = 500(1 - e^{-\ln(2 \times 300)/75})$$
$$\simeq 468.8 \text{ mg}.$$

This strategy shows that at the end of a 5-hour period the drug is almost cleared from the body.

• Exercises

1. A drug has half-time to elimination of 135 minutes. If a saturation level of 200 mg is to be maintained by 8-hourly doses of the drug (i.e. three times a day), what doses must be administered after the first one of 200 mg?

2. If the half-time of elimination for aspirin is 45 minutes, calculate the saturation level reached by a person taking 1 g of asprin every 2 hours (the recommended dose is 0.3–1.0 g every 4–6 hours).

3. Caffeine is the stimulatory drug present in many drinks, such as tea, coffee, and cola. Some facts about the drug follow.

A. It is more slowly absorbed than most drugs; in one hour about 25 per cent of the absorbed drug has been cleared.

B. A cup of coffee contains 100–50 mg of caffeine.

C. The convulsive dose of the drug is 10 g (i.e. the dose which can cause death).

(i) Calculate the approximate half-time of elimination for caffeine from the information above.

(ii) If a person drinks, on average, 10 cups of coffee per day (we shall assume them to be equally distributed throughout 24 hours), what will the saturation level of the caffeine in his body be?

(iii) Analyse the possibility of receiving a convulsive dose of caffeine by drinking coffee (one cup at a time!)

• Further reading

Doull, J., Klaasen, C. D., and Amdur, M. O. (1980). *The basic science of poisons*. Basingstoke, Macmillan.

Julien, R. M. (1978). *A primer of drug action*. San Francisco. W. H. Freeman & Co.

2.9. Population models

This section adopts a different style from the other sections in this book, and presents various models from population dynamics that illustrate the uses of ordinary differential equations.

The first census took place in 1801 and, with the exception of 1941, one has been conducted every 10 years. What is their purpose? The statistics derived from the census forms are used to predict future population and trends so that appropriate planning can take place. Are more hospitals required? Should colleges expand or contract? Should more motorways be built?

The idea of population prediction dates back to 1798 when Thomas Malthus wrote *An essay on the principle of population*.

THE MALTHUS POPULATION MODEL

What factors affect population growth? You could probably list four or five. In this first model we shall consider only three; the population itself $N(t)$, the birth rate, and the death rate.

Malthus argued that in a small time interval δt births and deaths are proportional to population size. Hence, in the small time δt, there are $\alpha N \delta t$ births and $\beta N \delta t$ deaths. The net increase in population δN is then

$$\delta N = \alpha N \delta t - \beta N \delta t = (\alpha - \beta) N \delta t$$

$$\frac{\delta N}{\delta t} = (\alpha - \beta) N = \gamma N.$$

γ, the growth rate, could be positive (more births than deaths) or negative (more deaths than births) or zero (equal numbers of births and deaths).

Letting $\delta t \to 0$ produces

$$\frac{dN}{dt} = \gamma N. \tag{2.14}$$

This is a first-order linear differential equation, and may be solved by separating the variables

$$\frac{1}{N}\frac{dN}{dt} = \gamma$$

$$\int \frac{1}{N} dN = \int \gamma\, dt = \gamma t + c$$

$$\ln N = \gamma t + c$$

$$N = e^{\gamma t + c} = e^{\gamma t} e^c = K e^{\gamma t}.$$

If the initial population is N_0, i.e. $N = N_0$ when $t = 0$, then $N_0 = K$ so

$$N(t) = N_0 e^{\gamma t}. \tag{2.15}$$

Figure 2.13 shows graphs of the population for positive, zero, and negative γ.

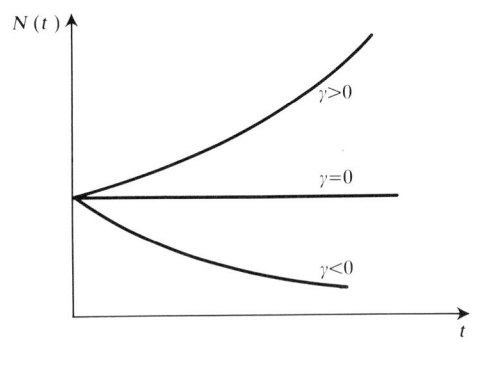

Fig. 2.13

The proof of the pudding is in the eating: does this population model describe what *actually* happens? The population of the USA for various years is listed in Table 2.1. The first two data points are used to determine the two unknown constants N_0 and γ and then predictions are calculated from eqn (2.15).

Table 2.1

Year	USA population ($\times 10^6$)	Malthusian model ($\times 10^6$)
1790	3.9	3.9
1800	5.3	5.3
1810	7.2	7.3
1820	9.6	10.0
1830	12.9	13.7
1840	17.1	18.7
1850	23.2	25.6
1860	31.4	35.0
1870	38.6	47.8
1880	50.2	65.5
1890	62.9	89.6
1900	76.0	122.5
1910	92.0	167.6
1920	106.5	229.3
1930	123.2	313.7

Calculus

$$t = 0 \quad \text{corresponds to} \quad 1790; \quad \text{thus} \quad N_0 = 3.9 \times 10^6.$$
$$t = 10 \quad \text{corresponds to} \quad 1800$$
$$N(10) = 5.3 \times 10^6 = 3.9 \times 10^6 \times e^{10\gamma}$$
$$\frac{5.3}{3.9} = e^{10\gamma}$$
$$\gamma = \frac{1}{10} \ln \frac{5.3}{3.9} = 0.0307.$$

Thus,
$$N(t) = 3.9 \times 10^6 \times e^{0.0307t}.$$

Any comments? The predicted values, although giving reasonable agreement at first become progressively too large. Why?

• Exercises

1. The birth rate of a species is 217 per 1000 per year, the death rate 211 per 1000 per year. What is the predicted population in (a) 1985; (b) 1990; (c) 2000 if the species numbered 3000 in 1980?

2. Suppose that in addition to births and deaths (with constant rates α and β, respectively) there is an increase in the population of a certain species due to migration of 100 individuals per unit time.

   (a) Formulate the equation for change in population δN.

   (b) Solve the equation.

   (c) What would you expect if $\alpha = \beta$? Does your solution agree?

3. In Exercise 1 how long will it take for the population to double?

THE VERHULST MODEL

The previous model predicted that the population will grow (or decay) exponentially. Although this seems to be fairly accurate in the initial stages, no population can grow exponentially indefinitely. What could prevent a population from growing without a bound?

Once a population grows sufficiently large it will begin to interact in a different way with its environment. Food, land, and energy will all limit growth, so that the growth rate will decrease as the population increases. This is known as the crowding factor. The growth rate γ is not a constant as in the Malthusian model, but depends on the population N. Thus

$$\frac{1}{N} \frac{dN}{dt} = \gamma(N). \qquad (2.16)$$

What can we say about $\gamma(N)$?

We have already reasoned that γ decreases as N increases. For very large populations the growth rate will be negative (this has been shown experimentally). The simplest curve with these properties is a straight line, as shown in Fig. 2.14.

N_∞ is the maximum population the environment can support without loss. Note that $N_\infty = a/b$, so

Calculus

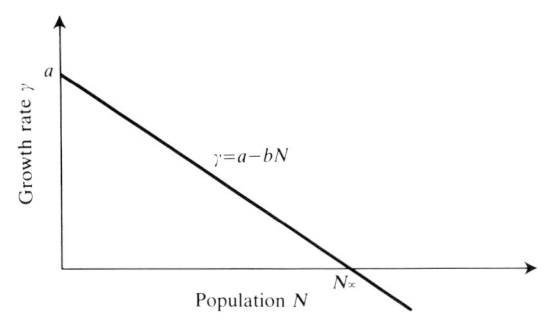

Fig. 2.14

$$\frac{1}{N}\frac{dN}{dt} = a - bN = a\left(1 - \frac{N}{N_\infty}\right).$$

This is called the logistic equation, and was developed by Verhulst in the 1830s. It is a first-order, nonlinear differential equation. It can be solved by separating variables and using partial fractions.

$$\int \frac{dN}{N(1 - N/N_\infty)} = \int a\,dt$$

$$\int \left[\frac{1}{N} + \frac{1/N_\infty}{(1 - N/N_\infty)}\right]dN = at + k$$

$$\ln\left(\frac{N}{1 - N/N_\infty}\right) = at + k.$$

This can be rearranged to give

$$N(t) = \frac{N_\infty}{1 + \left(\frac{N_\infty}{N_0} - 1\right)e^{-at}} \tag{2.17}$$

where $N(0) = N_0$. A typical curve of the solution in eqn (2.17) is illustrated in Fig. 2.15.

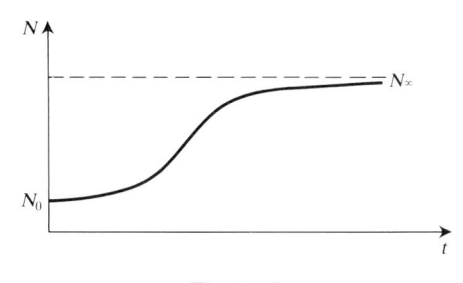

Fig. 2.15

Note that the population does not increase without limit as in the Malthus model. We will test the model using the USA population table; the results are shown in Table 2.2. The first three data points are used to determine the three unknown constants, N_0, N_∞, and a.

$$N_0 = 3.9 \times 10^6;$$
$$N_\infty = 197 \times 10^6;$$
$$a = 0.3134.$$

Calculus

Table 2.2

Year	Verhulst model ($\times 10^6$)	USA population ($\times 10^6$)
1790	3.9	3.9
1800	5.3	5.3
1810	7.2	7.2
1820	9.7	9.6
1830	13.0	12.9
1840	17.4	17.1
1850	23.0	23.2
1860	30.2	31.4
1870	38.1	38.6
1880	49.9	50.2
1890	62.4	62.9
1900	76.5	76.0
1910	91.6	92.0
1920	107.0	106.5
1930	122.0	123.0

Table 2.2 indicates an excellent fit—but is it good enough? After 1932 the Verhulst model goes astray. The population in the USA is now over 200×10^6 so a limiting population of 197×10^6 is obviously wrong.

- **Exercise**

Tables 2.3 and 2.4 show birth and death rates, respectively, in the UK. Formulate a Verhulst population model and compare your predictions with the actual population as illustrated in Table 2.5.

Table 2.3

Year	UK live births (per 1000 persons)
1971	16.1
1972	14.8
1973	13.8
1974	13.2

Table 2.4

Year	UK death rates (per 1000 persons)
1971	12.5
1972	12.6
1973	12.3
1974	11.9

Table 2.5

Year	UK population (in millions)	Year	UK population (in millions)
1971	55.610	1976	55.886
1972	55.781	1977	55.852
1973	55.913	1978	55.836
1974	55.922	1979	55.883
1975	55.900		

POPULATION MODEL WITH AGE DISTRIBUTION

To predict accurately future populations it is not sufficient to consider only the birth and death rates for the entire population. Consideration of birth and death rates for various age groups leads to a much clearer and more accurate population model. The age structure of a population can have wide variations as illustrated in Fig. 2.16.

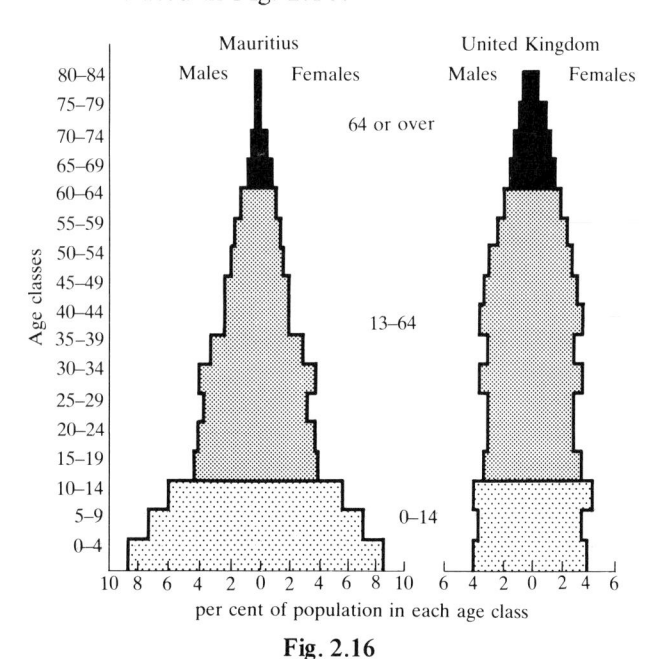

Fig. 2.16

Rather than use the 18 age groups shown in Fig. 2.16, we will simplify things and use only three age groups: 0–14, 15–44, and 45+. These groups have been chosen so that the birth rates in the first and last groups are zero, i.e. children under 14 and adults over 45 do not have children. Let $n_1(t)$, $n_2(t)$, and $n_3(t)$ denote the population in each group. We introduce a population *matrix* $N(t)$

$$N(t) = \begin{bmatrix} n_1(t) \\ n_2(t) \\ n_3(t) \end{bmatrix}.$$

b_i and d_i ($i = 1, 2, 3$) represent the birth and death rates, respectively, in each group.

We now formulate three recurrence equations, relating the population in year $t + 1$ to that in year t. For convenience, we assume an even distribution of ages in each group. Is this true for Mauritius?

Calculus

Consider group 1. After one year the population in group 1 comprises those born during that year, together with those of group 1 who have survived but not advanced to group 2.

$$n_1(t+1) = b_1 n_1 + b_2 n_2 + b_3 n_3 + \frac{14}{15}(1-d_1)n_1$$

$$n_1(t+1) = b_2 n_2 + \frac{14}{15}(1-d_1)n_1. \tag{2.18}$$

Consider group 2. After one year, the population will comprise those from group 1 who have survived and advanced to group 2 together with those of group 2 who have survived but not advanced to group 3.

$$n_2(t+1) = \frac{1}{15}(1-d_1)n_1 + \frac{29}{30}(1-d_2)n_2. \tag{2.19}$$

Similarly, for group 3,

$$n_3(t+1) = \frac{1}{30}(1-d_2)n_2 + (1-d_3)n_3. \tag{2.20}$$

Equations (2.18)–(2.20) can be represented in matrix form as follows.

$$\begin{bmatrix} n_1(t+1) \\ n_2(t+1) \\ n_3(t+1) \end{bmatrix} = \begin{bmatrix} \frac{14}{15}(1-d_1) & b_2 & 0 \\ \frac{1}{15}(1-d_1) & \frac{29}{30}(1-d_2) & 0 \\ 0 & \frac{1}{30}(1-d_2) & 1-d_3 \end{bmatrix} \begin{bmatrix} n_1(t) \\ n_2(t) \\ n_3(t) \end{bmatrix}$$

or

$$N(t+1) = AN(t)$$

where A is the *transition matrix*. A has 3 rows and 3 columns and so is described as a 3×3 matrix.

Now

$$N(1) = A\ N(0)$$
$$N(2) = A\ N(1) = A^2\ N(0)$$
$$N(3) = A\ N(2) = A^3\ N(0)$$

and, in general,

$$N(i) = A^i\ N(0)$$

To predict the population we need to know $N(0)$, the initial population, and d_1, d_2, d_3, b_2 so that the elements of the matrix A can be calculated. For the initial population we use the population of England and Wales in 1881.

$$N(0) = \begin{bmatrix} 9\,498 \\ 11\,641 \\ 4\,907 \end{bmatrix} \text{ using units of 1000 people.}$$

$b_1 = b_3 = 0; b_2 = 0.075; d_1 = 0.02; d_2 = 0.008; d_3 = 0.062.$

$$A = \begin{bmatrix} 0.915 & 0.075 & 0 \\ 0.065 & 0.959 & 0 \\ 0 & 0.033 & 0.938 \end{bmatrix}.$$

Thus,

$$N(1) = A\, N(0) = \begin{bmatrix} 0.915 & 0.075 & 0 \\ 0.065 & 0.959 & 0 \\ 0 & 0.033 & 0.938 \end{bmatrix} \begin{bmatrix} 9\,498 \\ 11\,641 \\ 4\,907 \end{bmatrix} = \begin{bmatrix} 9\,575 \\ 11\,781 \\ 4\,987 \end{bmatrix}.$$

The total population for 1882 is calculated as $26\,343 \times 10^3$. Similarly,

$$N(2) = \begin{bmatrix} 9\,645 \\ 11\,920 \\ 5\,067 \end{bmatrix} \quad N(3) = \begin{bmatrix} 9\,719 \\ 12\,058 \\ 5\,146 \end{bmatrix} \quad N(4) = \begin{bmatrix} 9\,797 \\ 12\,195 \\ 5\,225 \end{bmatrix}$$

$$N(5) = \begin{bmatrix} 9\,879 \\ 12\,332 \\ 5\,303 \end{bmatrix} \quad N(6) = \begin{bmatrix} 9\,964 \\ 12\,469 \\ 5\,381 \end{bmatrix}.$$

Table 2.6 compares these results with the actual population for the years 1881–7. The agreement is very good. In this model, birth and death rates have been taken as constant. If these rates altered, a new transition matrix would have to be calculated.

Table 2.6

Year	Predicted population (thousands)	Actual population (thousands)
1881	26 046	26 046
1882	26 343	26 334
1883	26 632	26 627
1884	26 923	26 922
1885	27 217	27 220
1886	27 514	27 522
1887	27 814	27 827

- **Exercises**

1. The population matrix for 1881 for the UK is

$$\begin{bmatrix} 9\,900 \\ 12\,140 \\ 5\,400 \end{bmatrix}$$

Calculus

Using A as given verify that

$$A^{10} = \begin{bmatrix} 0.537 & 0.449 & 0 \\ 0.389 & 0.801 & 0 \\ 0.0591 & 0.218 & 0.527 \end{bmatrix}.$$

(Hint: calculate A^2, A^4, and A^8.) Hence estimate the 1891 population for the UK. How does this compare with the actual value?

2. A 1976 census produced an age profile as illustrated in Table 2.7. By simplifying the information into three age groups make a forecast for the population in the year 2000.

Table 2.7

Age group	1976 age distribution (in millions)	
	Males	Females
Under 15	6.7	6.4
15–29	6.2	6.0
30–44	5.0	4.9
45–64	6.3	5.0
65–74	2.2	4.6
Over 75	0.9	1.9

3. Mechanics

Mechanics

3.1. Pursuit curves

• Differential equations; numerical methods

Guided missiles have various systems for homing on their target. One method is for the missile to use the exhaust heat of an aeroplane to find the position of the aircraft. The missile then moves towards this heat source. Thus the missile is always pointing towards its target. In this section we find the missile's path and also the time it takes to capture the aircraft.

• Analytical solution

Suppose that at time $t = 0$ the missile is at the origin $(0, 0)$ and the aeroplane is at (a, b). Also, suppose that the aeroplane moves parallel to the x-axis with constant velocity v_a and the missile with constant speed v_m. Suppose further that at time t the position of the aeroplane is (x_A, b) and the position of the missile is (x_m, y_m) (see Fig. 3.1). The tangent at M to the missile's path passes through A. The equation of the tangent is

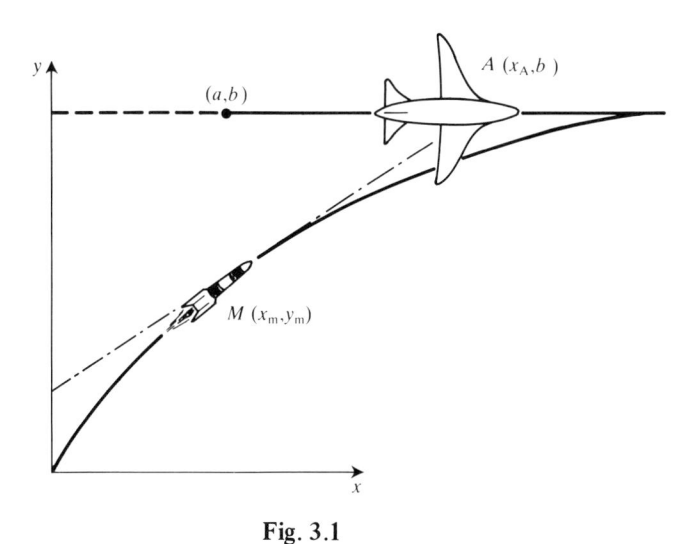

$$y - y_m = \left(\frac{dy}{dx}\right)_m (x - x_m)$$

$$y - y_m = \left(\frac{dy}{dt} \Big/ \frac{dx}{dt}\right)_m (x - x_m) \ldots \qquad (3.1)$$

Fig. 3.1

But $A(x_A, b)$ lies on this tangent and

$$x_A = a + v_A t.$$

From eqn (3.1), $b - y_m = (\dot{y}_m/\dot{x}_m)(a + v_A t - x_m)$ and $v_m^2 = \dot{x}_m^2 + \dot{y}_m^2$ (writing $dx/dt = \dot{x}$ and $dy/dt = \dot{y}$). Thus the path of the missile is given by the equations

$$\dot{x}(b - y) = \dot{y}(a + v_A t - x) \qquad (3.2)$$

$$\dot{x}^2 + \dot{y}^2 = v_m^2 \qquad (3.3)$$

after dropping the suffix m.

The initial conditions are $x(0) = y(0) = 0$. Now eqn (3.2) may be written as

Mechanics

$$\frac{dx}{dy}(b-y) = a + v_A t - x. \tag{3.2a}$$

Differentiating with respect to t,

$$\frac{d^2x}{dy^2}\frac{dy}{dt}(b-y) - \frac{dx}{dy}\frac{dy}{dt} = v_A - \frac{dx}{dt},$$

so that

$$\frac{d^2x}{dy^2}\frac{dy}{dt}(b-y) = v_A. \tag{3.4}$$

From (3.3)
$$\dot{y}^2(1 + (\dot{x}/\dot{y})^2) = v_m^2,$$

$$\frac{dy}{dt} = v_m \bigg/ \left\{1 + \left(\frac{dx}{dy}\right)^2\right\}^{1/2}. \tag{3.5}$$

Substituting for dy/dt from eqn (3.5) into eqn (3.4),

$$\frac{d^2x}{dy^2}(b-y) = k\left\{1 + \left(\frac{dx}{dy}\right)^2\right\}^{1/2} \tag{3.6}$$

where $k = v_A/v_m$. This is a second-order, non-linear differential equation.

Let $p = dx/dy$, so that $dp/dy = d^2x/dy^2$. Then eqn (3.6) may be written

$$\frac{dp}{dy}(b-y) = k(1+p^2)^{1/2},$$

i.e. $\ln\{p + (1+p^2)^{1/2}\} = -k\ln(b-y) + A$

where A is constant.

Now, when $t = 0$, $y = 0$, and $p = dx/dy = a/b = c$, say,

$$A = \ln\{c + (1+c^2)^{1/2}\} + k\ln b$$
$$= \ln(db^k)$$

where $d = c + (1+c^2)^{1/2}$. Thus,

$$\ln\{p + (1+p^2)^{1/2} = \ln(b-y)^{-k} + \ln(db)^k.$$

Therefore,

$$p + (1+p^2)^{1/2} = \frac{db^k}{(b-y)^k}$$

$$(1+p^2)^{1/2} = \frac{db^k}{(b-y)^k} - p.$$

Squaring,

$$1 + p^2 = \frac{d^2 b^{2k}}{(b-y)^{2k}} - \frac{2pdb^k}{(b-y)^k} + p^2$$

$$p = \left\{\frac{d^2 b^{2k}}{(b-y)^{2k}} - 1\right\}\frac{(b-y)^k}{2db^k}.$$

Replacing p by dx/dy,

$$\frac{dx}{dy} = \frac{1}{2}\frac{b^k}{(b-y)^k} - \frac{(b-y)^k}{b^k}.$$

Integrating with respect to y,

62

$$x = \frac{1}{2}\left\{\frac{db^k}{(k-1)(b-y)^{k-1}} + \frac{(b-y)^{k+1}}{(k+1)db^k}\right\} + B$$

for some constant B. Here we are assuming that $k \neq 1$, since we must assume

$$v_m > v_A, \quad \text{i.e.} \quad k < 1.$$

We have initial conditions $x = y = 0$, giving

$$B = \frac{1}{2}\left\{\frac{db}{1-k} - \frac{b}{d(k+1)}\right\}$$

$$= \frac{b}{2d(1-k^2)}\{d^2(1+k) - (1-k)\}$$

$$= \frac{b}{2d(1-k^2)}\{k(d^2+1) + d^2 - 1\}$$

Thus the missile's path is given by

$$x = \frac{1}{2}\left\{\frac{db^k(b-y)^{1-k}}{k-1} + \frac{(b-y)^{k+1}}{(k+1)db^k}\right\} + B. \tag{3.7}$$

The missile hits the target when $x = x_A = a + v_A t$ and $y = b$. Substituting in (3.7) with $y = b$, we get

$$a + v_A t = B,$$

i.e.

$$t = \frac{B-a}{v_A}.$$

Substituting for d and B and after some manipulation we get

$$t = \frac{(a^2+b^2)^{1/2} + ak}{v_m(1-k^2)}, \qquad k = \frac{v_A}{v_m}. \tag{3.8}$$

- **Numerical solution**

Suppose that at time t the aircraft is at (x_A, b) and the missile is at (x_m, y_m). After a further time Δt the aircraft is at (x'_A, b) and the missile is at (x'_m, y'_m). Then from Fig. 3.2 we have

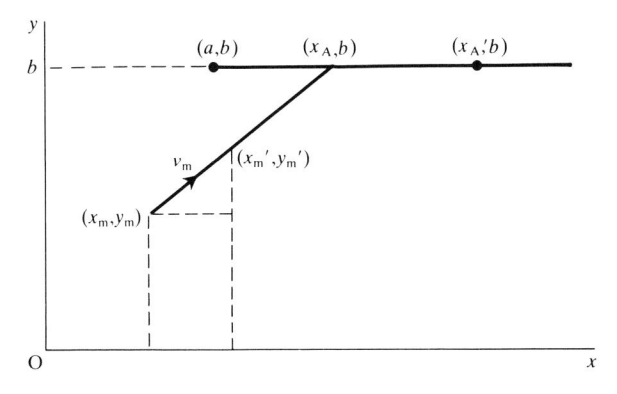

$$x'_A = x_A + v_A \Delta t,$$

$$x'_m = x_m + v_m \Delta t \cos\theta,$$

Fig. 3.2

Mechanics

and

$$y'_m = y_m + v_m \Delta t \sin \theta$$

and $t' = t + \Delta t$ where $\tan \theta = (b - y_m)/(x_A - x_m)$. These equations can be solved iteratively until $x'_m = x'_A$ and $y'_m = b$.

• Exercises

1. Use eqn (3.8) to show that, if the aircraft starts at $a = 2500$ m, $b = 3000$ m and $v_A = 1000$ m s^{-1}, $v_m = 2000$ m s^{-1}, then the missile hits the aircraft after 3.4 s.

2. Write a computer program to solve the problem iteratively. Care needs to be taken when deciding whether the missile has hit the aircraft.

3. A dog spots a cat at a distance a away. Suppose that the dog D is at $(0, a)$ and the cat C is at the origin $(0, 0)$. The cat is moving at speed v along the x-axis. What will be the path of the dog if its speed is w and it tries to catch the cat? Consider the cases $v < w$, $v = w$, and $v > w$.

Mechanics

3.2. Modelling river flow

• **Vectors; modelling**

From early experiences of dropping sticks into a river and watching their passage downstream, or later ones of swimming or boating in a river or estuary, it becomes evident that the speed at which the water flows may vary considerably across its width. For a straight stretch of river the flow will usually be fastest in the middle and tail away to almost zero at the banks, whilst on a bend the speed of flow is usually at its highest near the outside of the bend, where it erodes the bank, and slowest on the inside of the bend where it builds up deposits. Experienced boatmen make use of this knowledge to their advantage, taking the fast water when going downstream and seeking out the slow water when travelling upstream. Water authorities, too, need to have specialized knowledge of a river's cross-section and flow pattern to be able to make accurate estimates of the volume flow of water. Formulate a model for the velocity distribution across a river of width 20 m where the speed at the centre is 3 m s^{-1} (Fig. 3.3).

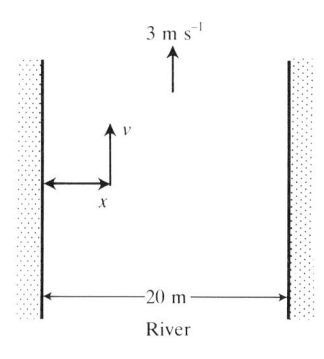

Fig. 3.3

• **Solution**

To take into account all the factors involved in the stream would be too complicated so we start with the simple situation of a long straight stretch of river with no boulders or obstacles, whose parallel banks are 20 m apart and where the stream flow is 3 m s^{-1} (Fig. 3.3) at the middle, falling off symmetrically to zero flow at the banks.

This does not really give us sufficient information to decide what the speed v of the water is at a point distance x from the left bank. We know that $0 \leqslant v \leqslant 3$ and that the speed increases from the banks towards the middle. Ideally we now need some experimental evidence to give us the river speed for different distances from the bank but without it we speculate about different possibilities and look at some of the consequences.

The essential ingredient of this aspect of mathematical modelling is curve-fitting. That is, we call on our experience of a variety of mathematical functions which will fit the known facts and then hopefully we will be able to use these to help predict something new about the situation.

Here we are trying to find $v = f(x)$, such that $f(0) = 0 = f(20)$ and $f(10) = 3$ with symmetry about $x = 10$ and such that $0 \leqslant f(x) \leqslant 3$ for $0 \leqslant x \leqslant 20$. This may seem overwhelming at first but a few examples should soon clarify the situation.

Mechanics

The most logical model to start with is that which assumes the river's speed increases at a constant rate from each bank to the middle as shown in Fig. 3.4. This leads to the function

$$v = \frac{3}{10}x, \qquad 0 \leqslant x \leqslant 10$$

$$= -\frac{3}{10}x + 6, \qquad 10 \leqslant x \leqslant 20.$$

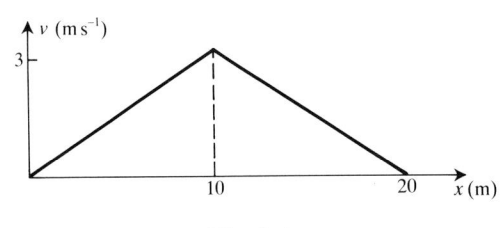

Fig. 3.4

In the succeeding models we make use of our experience which suggests that the speed of a river drops off slowly from its peak near the middle giving a rounded curve for the graph rather than the sharp point at $x = 10$. Here we take the unique quadratic which fits the facts (see Fig. 3.5). This must have the form $v = kx(x - 20)$ because of where it cuts the axis and, as $v = 3$ when $x = 10$, then $k = -3/100$, giving

$$v = 3x(20 - x)/100.$$

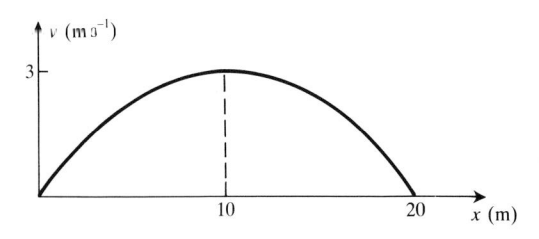

Fig. 3.5

Another function which has the right kind of symmetry is the sine function, (see Fig. 3.6). At first glance it does not look much different to the previous model but the sine curve is more pointed than the parabolic curve of that model. Which curve is the more appropriate for a given situation will depend on the river's cross-section. In this model,

$$v = a \sin bx$$

and, to fit the known facts, $bx = \pi$ when $x = 20$, while $v = 3 = a$ when $x = 10$. This gives

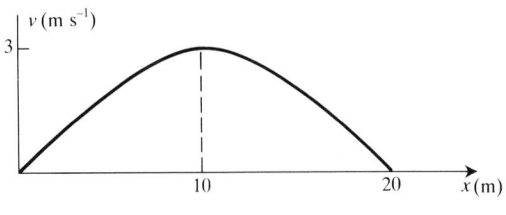

$$v = 3 \sin \frac{\pi}{20}x.$$

Fig. 3.6

Mechanics

A fourth function which may be the best model for many situations is that corresponding to an ellipse as shown in Fig. 3.7. To obtain the equation, write down the standard equation for an ellipse with semi-axes of 10 and 3 and then translate 10 units to the right. This leads to

$$v = \frac{3}{10}[x(20-x)]^{\frac{1}{2}}.$$

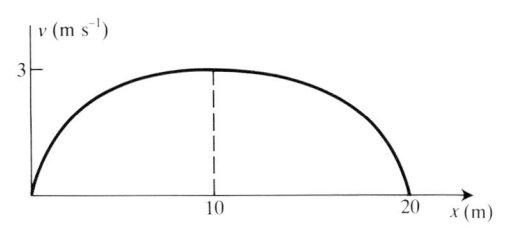

Fig. 3.7

Models 1 to 4 have become mathematically more sophisticated as we have tried to find functions which give a better fit to the known data and our own experience of flow patterns. In practice, however, the mathematically sophisticated functions may be difficult to handle and an approach to approximating to the data by a sequence of straight lines may be adequate. Model 5 (see Fig. 3.8) illustrates the approach as, in fact, does Model 1.

In the case of Model 5

$$v = \frac{3}{5}x, \qquad 0 \leqslant x \leqslant 5$$

$$= 3, \qquad 5 < x < 15$$

$$= -\frac{3}{5}x + 12, \qquad 15 \leqslant x \leqslant 20.$$

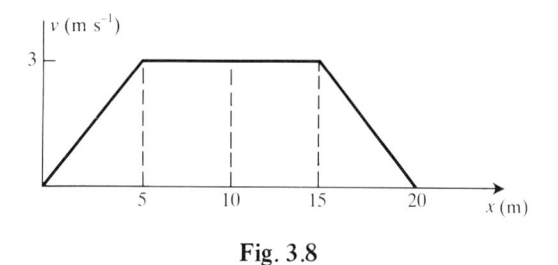

Fig. 3.8

• Worked example

If you have ever observed a boat or ferry crossing a river, you may often have noticed what an erratic course it takes and it is this which we now investigate, making use of one of the models developed above.

From experience, a ferryman knows that by starting at a point A on the left bank (Fig. 3.9) and heading his boat at an angle upstream with velocity $4\mathbf{i} - 2\mathbf{j}$ relative to the water (where $\mathbf{i}$ is a unit vector across the stream, and $\mathbf{j}$ a unit vector parallel to the river bank), he will end up at the point B immediately opposite A. (The units of speed here are metres per second and $\mathbf{i}$ and $\mathbf{j}$ are as in Fig. 3.9.)

Mechanics

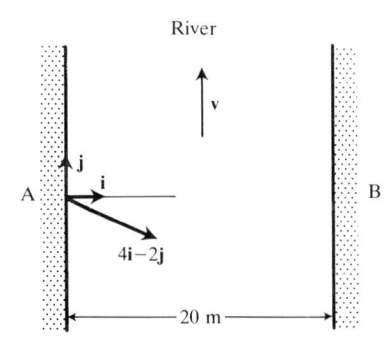

Fig. 3.9

We use the Model 2 for the water's velocity and this gives the boat's velocity relative to an observer on the bank as

$$\mathbf{v}_B = 4\mathbf{i} + \left[\frac{3}{100} \times (20 - x) - 2\right]\mathbf{j}.$$

This velocity has two components, one of 4 m s^{-1} across the stream in the $\mathbf{i}$ direction and one parallel to the bank in the $\mathbf{j}$ direction which is a function of its distance x from the bank. Now, if t is the time in seconds after the ferry leaves A, then

$$x = 4t$$

so

$$\mathbf{v}_B = 4\mathbf{i} + \left[\frac{12}{25}t(5 - t) - 2\right]\mathbf{j}.$$

To find where the boat is at time t we need to integrate the expression for the velocity to obtain the boat's position vector

$$\mathbf{r} = 4t\mathbf{i} + \left(\frac{6}{5}t^2 - \frac{4}{25}t^3 - 2t\right)\mathbf{j} + \mathbf{c}.$$

Taking A as the origin, then $\mathbf{c} = 0$, and factorizing the cubic in t gives

$$\mathbf{r} = 4t\mathbf{i} - \frac{2}{25}t(2t - 5)(t - 5)\mathbf{j}.$$

Finding $\mathbf{r}$ for different values of t in the range $0 \leqslant t \leqslant 5$ enables the path of the boat to be plotted as in Fig. 3.10.

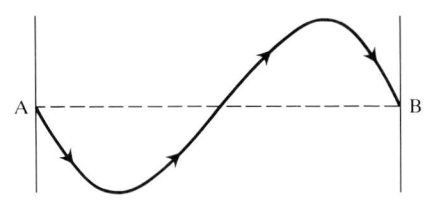

Fig. 3.10

• Exercises

1. For Models 1–3 investigate the path of a rowing boat which leaves the left bank at A and is rowed at a steady speed of 1 m s^{-1} at right angles to the bank across the river.

    For which model does the boat go furthest downstream?

Mechanics

2. On a bend in the river the highest speed of the water is 4 m s^{-1} and occurs 5 m from the outside bank of the bend. Assuming that the water speed is zero at the banks and that the river is 20 m wide, find possible functions to model the water speed x metres from the bank.

3. The above models have all been concerned with the water speed near the surface of the river. To obtain an estimate of the volume-flow of water along a river, we need to know its cross-section and the speed of water at all points of the cross-section.

   As an example of such a problem, consider a river whose vertical cross-section can be approximated by a semicircle radius 10 m (see Fig. 3.11). Further, suppose the water speed at all points distance r from the centre 0 is given by

   $$v = (10 - r)^2/50 \text{ m s}^{-1}.$$

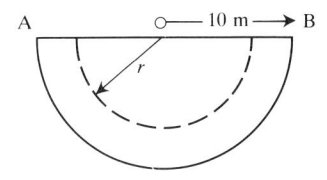

Fig. 3.11

What is the highest speed predicted? Find the rate of flow of water along the river in cubic metres per second. What water speed, assumed constant over all the cross-section, would give the same volume flow?

Mechanics

3.3 The tennis service

• **Projectiles**

Wimbledon watchers and tennis players cannot fail to be aware of the importance of the service. Many professional players are noted for the high speed of their service, whilst the fate of a championship often depends on which player manages to get the larger proportion of his first service into play. Why is it so difficult to get a service in play when the region in which the ball is permitted to land is a rectangle 21 feet long by $13\frac{1}{2}$ feet wide? Why does a server who fails to get this service in usually reduce the speed of his second attempt? Can we arrive at a better understanding of the situation?

• **Solution**

Model 1

There are many factors which influence the flight of a tennis ball. Its speed, gravity, the ball's surface, wind resistance, the effect of spin, humidity, temperature, to name but a few. However, much can be gained by ignoring all these factors initially and we just assume that the ball travels in a straight line. This may seem a crude approximation but for a fast service it is not very different from what we observe. This assumption is equivalent to ignoring all the forces on the ball or assuming that their effect is negligible for the time taken for the service.

Suppose the ball is struck by the server at D (see Fig. 3.12), at a height h above the centre A of the base line. Where can the ball land in the receiver's court? Here we make a simplifying assumption, that the ball is a point particle, and further we assume that the top of the net is a straight line 3 feet above the middle of the court. If the ball is hit in the plane BAD to just clear the net, then if x is the distance from the net where the ball first strikes the ground at P, we have (see Fig. 3.13)

$$\frac{x}{3} = \frac{x+39}{h},$$

giving

$$x = \frac{117}{h-3}.$$

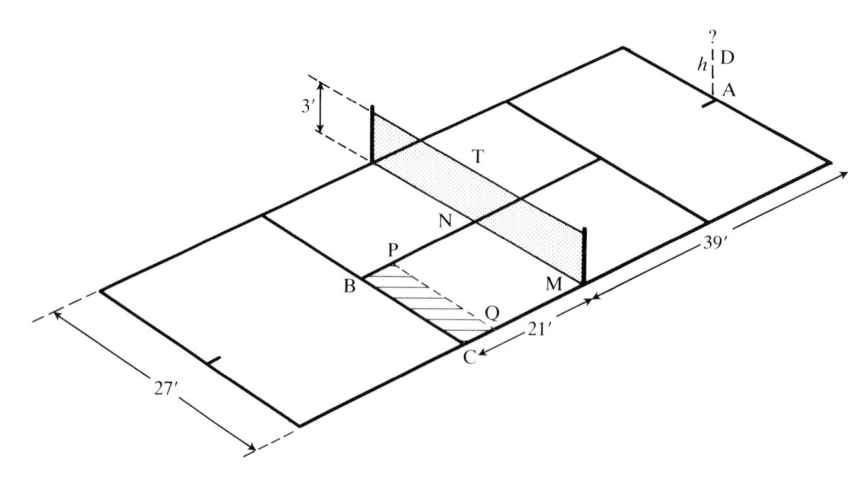

Fig. 3.12

Mechanics

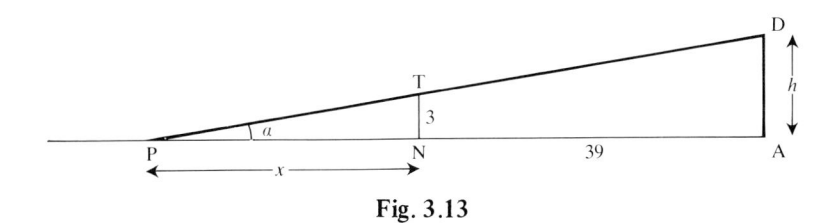

Fig. 3.13

The point at which the ball strikes the ground is thus dependent on the height at which it was struck. A typical height at which a man strikes the ball for a service is about 9 feet and this gives

$$x = \frac{117}{6} = 19.5 \text{ feet.}$$

But the edge of the service box is only 21 feet from the net, so this allows the server very little room to manoeuvre.

What would be the effect of the server hitting the ball to just clear the net at some angle to the centre line? The neatest way to see the answer is to observe that the paths of all the balls struck from D to just clear the net lie in the plane formed by D and the line of the top of the net. This plane intersects the plane of the court in a straight line PQ parallel to the net. When $h = 9$ feet, the line PQ is 19.5 feet from the net so the only area into which the server can hit the ball is the narrow strip BPQC (see Fig. 3.12), which is $1\frac{1}{2}$ feet wide and represents only 1/14 of the area of the service box.

This model has made many simplifying assumptions but for the Tanners of the tennis world, whose high-speed service approximates to a straight line, it highlights their problem of essentially what a small target area they have to aim at. A tall player could have a distinct advantage. If a player could serve from a height of 10 feet, then $x \simeq 16.7$ feet, giving him a target strip almost three times as wide as the player who serves from a height of 9 feet.

On the other hand it suggests why the shorter tennis players could never employ a high-speed service. Suppose such a player can serve from a height of 8 feet; then this model gives

$$x = 117/5 > 21 \text{ feet}$$

and so would always be served long.

Perhaps there is a message here for all budding tennis players—hit the ball from as high a point as you can to increase your target area.

Before looking at a better model for a slow service, we briefly consider the range of angles (Fig. 3.14) at which the first server can hit a ball from a height of 9 feet and get an ace. At one extreme the ball will land at B (θ_1) and at the other (θ_2) it will just clear the net and land at P.

$$\theta_1 = \tan^{-1}\frac{9}{60} \simeq 8.53° \qquad \theta_2 = \tan^{-1}\frac{9}{58.5} \simeq 8.75°.$$

Thus

$$8.53° < \theta < 8.75°$$

is the angular region in which the server can get an ace. Such a small margin of error $\pm 0.1°$ at which the ball can be struck makes it very easy to see why a fast server has so many first services which are called long or hit the net cord.

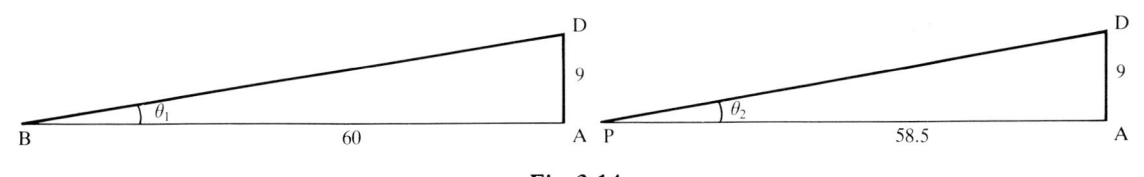

Fig. 3.14

Mechanics

Model 2

For a better model of the flight of a tennis ball we will take the acceleration due to gravity into account but still ignore all the aerodynamic effects. The ball is then considered as a point particle projected with speed v at an angle α to the horizontal with the point of projection taken as the origin (see Fig. 3.15).

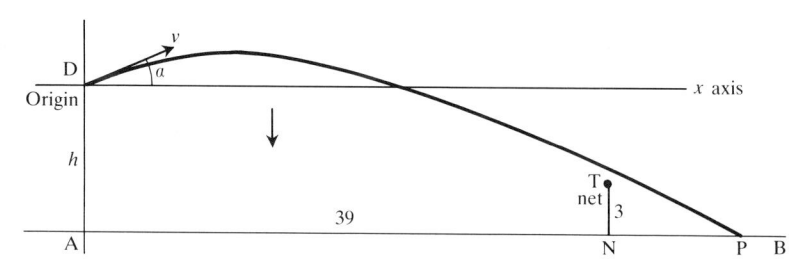

Fig. 3.15

After a time t the horizontal and vertical components of the ball's position will be given by

$$y = vt \cos\alpha$$
$$y = vt \sin\alpha - \tfrac{1}{2}gt^2$$

using

$$s = ut + \tfrac{1}{2}at^2.$$

Eliminating t gives the equation of the ball's trajectory as

$$y = x\tan\alpha - \frac{gx^2}{2v^2}\sec^2\alpha.$$

To find where the ball can land we need to see how close to the base of the net P comes if the ball just clears the net at T. This is equivalent to the trajectory of the ball going through the point $T(39, -(h-3))$ which leads to the relation

$$3 - h = 39\tan\alpha - \frac{39^2 g}{2v^2}\sec^2\alpha,$$

involving three variables over which the server has some control; the height h from which he serves, the speed v, and the angle α.

By writing $\sec^2\alpha = 1 + \tan^2\alpha$ in the last equation it can be rearranged as a quadratic equation in $\tan\alpha$ showing that for given values of h and v there are two angles of projection for the service to just clear the net (see Fig. 3.16).

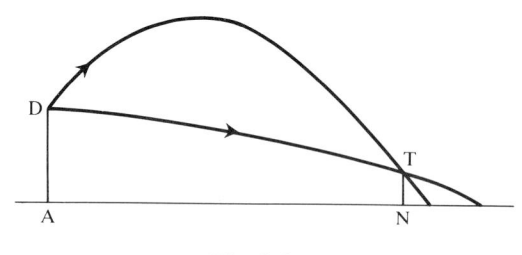

Fig. 3.16

In practice the server usually uses the lower trajectory, although the higher trajectory would allow him to exploit more of his opponent's service box.

What is the advantage of the lower trajectory?

When v and α are known, the point at which the ball strikes the court is given by putting $y = -h$ in the equation of the trajectory and solving the resulting quadratic equation in x. Why are there two values for x and which one has to be discarded?

• Exercises

1. Use the first model to determine the minimum height from which a fast service could be hit and still be served in.

2. How would you modify the first model to take into account the physical size of a tennis ball?

3. A tennis player always serves so that the ball is projected horizontally from a height of 9 feet.

   Use the second model to find the speed at which he serves if the ball

   (i) just clears the net; (ii) strikes the court at B.

4. Tennis services have been measured at over 150 m.p.h. (220 ft/s). Use the second model to find the angle at which a service has to be hit from a height of 9 feet to just clear the net and hence find where it hits the court. How does this compare with the prediction of the first model?

• Further reading

To improve on the models discussed here some account must be taken of the aerodynamic effects of the resistance and spin. The effect of air resistance is discussed in most textbooks on A-level dynamics, whilst the effect of spin is discussed in

Daish, C. B. (1972). *The physics of ball games.* English Universities Press.

Mechanics

3.4. Head-on crash

• **Motion with constant acceleration; impulse; momentum**

Nearly 50 per cent of all road accidents involve at least one of the vehicles concerned in a head-on collision either with another vehicle, stationary or moving, or a fixed object such as a lamp-post or a wall. In this section we investigate the effect on passengers in the vehicle and particularly the differences brought about by wearing seat-belts.

• **Solution**

How does a car behave in an impact with a solid object such as a wall?

Most cars these days are designed so that the passenger compartment is well protected but the engine and boot space will crumple in a crash.

Why is this? When the car is travelling at an initial velocity u(m/s) it has to slow down during the impact to 0 m/s. Suppose that during the impact the front d metres of the bonnet collapses and the passenger compartment remains intact (see Fig. 3.17). Then we can estimate the deceleration during the crash. Experimental results show that the graph of deceleration against time is shown in Fig. 3.18. However we obtain a theoretical model assuming a constant deceleration a(m s^{-2}).

Fig. 3.17

Using $v^2 - u^2 = 2as$ with $v = 0$ when $s = d$ we get

$$a = -u^2/2d.$$

Typically d would be between 0.5 and 2 m so, taking d to be an average value of, say, 1.2 m, the deceleration due to impacts at various speeds would be as shown in Table 3.1.

The calculation is as follows. For example,

$$30 \text{ m.p.h.} = \frac{30 \times 1609.3}{3600} = 13.4 \text{ m/s}.$$

74

Mechanics

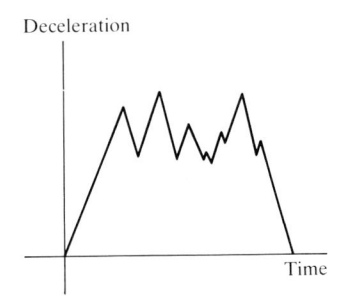

Fig. 3.18

Table 3.1

Speed (m.p.h.)	10	20	30	40	50	60	70
Deceleration (g)	0.85*	3.4*	7.6	13.6	21.2	30.5	41.6

*These figures are suspect as these speeds may not produce complete collapse of the bonnet.

Hence,

$$a = \frac{u^2}{2d} = \frac{13.4^2}{2 \times 1.2} = 74.8 \text{ m/s}^2.$$

This is approximately $7.6g$, i.e. nearly 8 times the acceleration due to gravity. Notice that this is an average value (in fact, the mean value) and the peak could well be twice this, from the form of the graph in Fig. 3.18.

What difference does it make if the collision is with a stationary vehicle rather than a rigid obstruction?

The figures given above are for a collision with immovable, rigid objects or, equivalently with a vehicle of similar mass and construction travelling in the opposite direction with a similar speed. If, instead, the collision is with a stationary vehicle of similar mass and construction there are three mitigating effects:

1. The distance over which deceleration takes place will be greater because the car impacted on will have a collapse zone;
2. The change in velocity will be less because the two vehicles will have some (common?) velocity after the impact;
3. The distance over which deceleration takes place will be greater because the car impacted will start to move during the deceleration phase.

Let us consider the impact; other problems may arise afterwards, but we will not concern ourselves with them. If the situation immediately before and after the impacts is as shown in Fig. 3.19 and the masses of the cars are similar at m(kg), then, by conservation of momentum,

$$mu = 2mv, \quad \text{i.e.} \quad v = \tfrac{1}{2}u.$$

Suppose that a constant, and common, force F (newtons) is required to compress the collapse zones of the two cars; then the deceleration/acceleration will split into two parts, the collapse phase and a sudden impulse as the rigid passenger compartments meet. Obviously it is desirable for the sake of the occupants that this sudden impulse should be minimal!

By the end of the collapse phase let the velocities of the two cars be v_1 and v_2; then we can

Mechanics

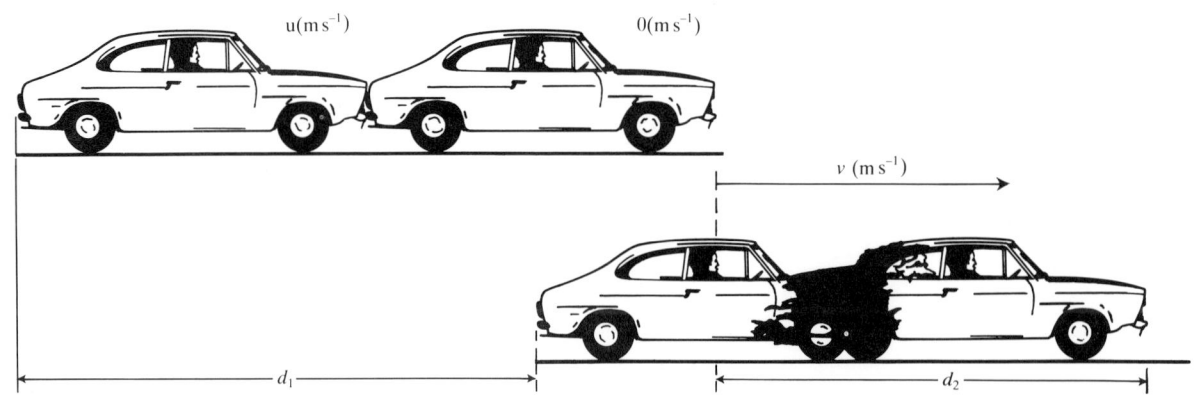

Fig. 3.19

form the following equations, where c_1 and c_2 are the lengths of the front and rear collapse zones and t the duration of the collapse phase,

$$d_1 - d_2 = c_1 + c_2 \tag{3.9}$$

where d_1 and d_2 are the distances moved by the uncollapsed portions of the two cars. From the impulse/momentum equation,

$$Ft = m(u - v_1) \tag{3.10}$$
$$= mv_2. \tag{3.11}$$

From the equations of motion with constant acceleration,

$$d_1 = \tfrac{1}{2}(u + v_1)t \tag{3.12}$$
$$d_2 = \tfrac{1}{2}v_2 t \tag{3.13}$$

From equs (3.12) and (3.13)

$$\frac{d_1}{d_2} = \frac{u + v_1}{v_2}. \tag{3.14}$$

Equating the right-hand sides of (3.10) and (3.11) gives

$$u - v_1 = v_2 \tag{3.15}$$

and substituting in (3.14) gives

$$\frac{d_1}{d_2} = \frac{u + v_1}{u - v_2},$$

i.e. $d_1(u - v_1) = d_2(u + v_1)$. Now from (3.9),

$$d_1 = c_1 + c_2 + d_2$$

so

$$(c_1 + c_2 + d_2)(u - v_1) = d_2(u + v_1)$$

i.e.

$$(c_1 + c_2)(u - v_1) = d_2(u + v_1) - d_2(u - v_1)$$
$$= d_2(u + v_1) - (u - v_1)$$
$$= 2d_2 v_1. \tag{3.16}$$

If we now suppose $v_1 = v_2$ so that there is no sudden impulse when passenger compartments meet, then $v_1 = v_2 = \tfrac{1}{2}u$ from (3.15).

Typically, $c_1 = 1.2$ m, $c_2 = 0.8$ m, $m = 900$ kg. Now from (3.16) with $v_1 = \frac{1}{2}u$, $c_1 = 1.2$, $c_2 = 0.8$, we get

$$d_2 = \frac{(1.2 + 0.8)(u - \frac{1}{2}u)}{2 \cdot \frac{1}{2}u} = 1$$

Thus $d_2 = 1$ m whatever the impact speed!

Eqn (3.14) gives $\quad d_1 = 3m$,

Eqn (3.13) gives $\quad t = \dfrac{2d_2}{v_2} = \dfrac{4}{u}$,

Eqn (3.11) gives $\quad F = \dfrac{mv_2}{t} = \frac{1}{4}u \times m \times \frac{1}{2}u = \frac{1}{8}mu^2$.

As F depends on u, we cannot manufacture cars to cope with any impact speed. If we give F a value sufficiently large to cope with the maximum likely impact speed, say 60 m.p.h., then for slower impacts the full collapse will not take place and the full benefit of the collapse zones will be lost. Now the value of F for 60 m.p.h., i.e. 26.8 m s^{-1}, is given by $F = \frac{1}{8} \times 900 \times 26.8^2 = 8.1 \times 10^4$ Newtons (i.e. 8.1 tonnes wt. approx.), whereas the value required for 30 m.p.h. will be only a quarter of this since $F \propto u_2$.

This explains why in practice the collapse zones are built to require ever-increasing force to collapse them rather than a constant value. This ensures that the velocities of the two cars are equalized without a sudden impulse, whatever the impact velocity, while at the same time retaining the majority of the benefit imparted by the collapse zones during impacts at low speeds.

Going back to the assumption of a constant force, as a comparison the acceleration during the collapse is $u^2/8$ as opposed to $u^2/2.4$ for impact with solid objects.

After the collapse further deceleration is due to friction (high, due to battered bodywork!) and any subsequent impacts, but these will be less drastic as the initial velocity has already been halved.

How is the passenger in the car affected by this deceleration?

What the passengers (or driver) (Fig. 3.20) experiences is the windscreen acceleration towards them, if they are not wearing seatbelts. The speed with which it hits them will depend upon their initial distance from the screen and its acceleration towards them, which in turn depends upon the impact velocity. If this is a stationary vehicle, the speed with which the passenger hits the windscreen (or vice versa, depending on your point of view) can be found using $v^2 - u^2 = 2as$ with $u = 0$ (since the initial velocity of the windscreen relative to the passenger is zero). Thus $a = \frac{1}{4}u^2$, $s = d$ (where d is the passenger's initial distance from the windscreen) gives $v^2 = \frac{1}{4}u^2 d$, i.e. $v = \frac{1}{8}u\sqrt{d}$.

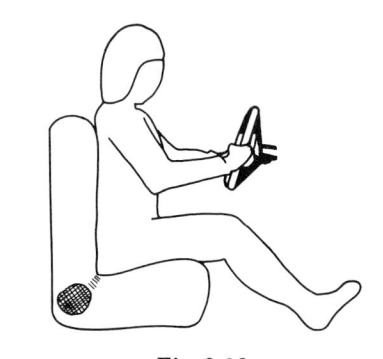

Fig. 3.20

Mechanics

It is worth noting that if $d = 0$, i.e. passengers are sitting with their noses pressed to the screen, then there is no impact on the screen ($v = 0$). Whether the passengers go through the screen is then determined by whether it can provide the force necessary to decelerate the passengers at the same rate as the car without breaking. This force is given by

$$F = ma,$$

i.e.

$$F = m_p \tfrac{1}{8} u^2 = \tfrac{1}{8} m_p u^2$$

where m_p is the mass of the passengers.

Consider a 10 stone man in a 40 m.p.h. collision. Converting this to S.I. units this force is

$$\tfrac{1}{8} \times 140 \times 0.45 \times \left(\frac{(40 \times 1609)^2}{3600} \right) = 2517 \text{ newtons.}$$

The crucial factor is likely to be the area over which this is spread. Graphically this situation is shown in Fig. 3.21.

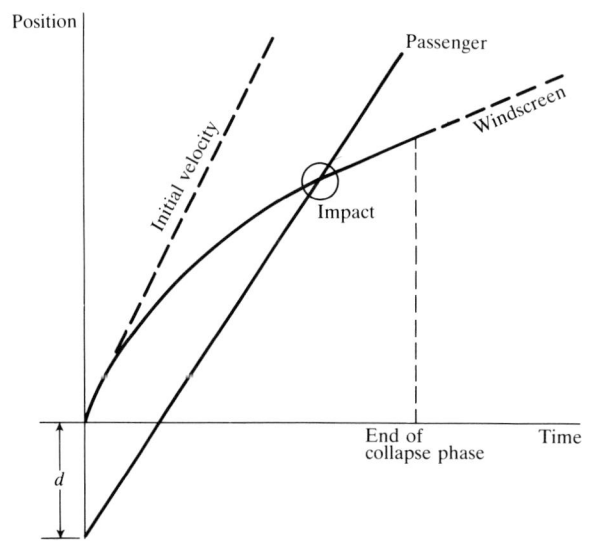

Fig. 3.21

How does a seat-belt help?

Suppose that, after an initial amount of 'give', the seat-belt becomes taut* and provides sufficient force to decelerate the passenger at the same rate as the car. The graph in Fig. 3.21 will then be modified to look like Fig. 3.22. This has obvious beneficial results!

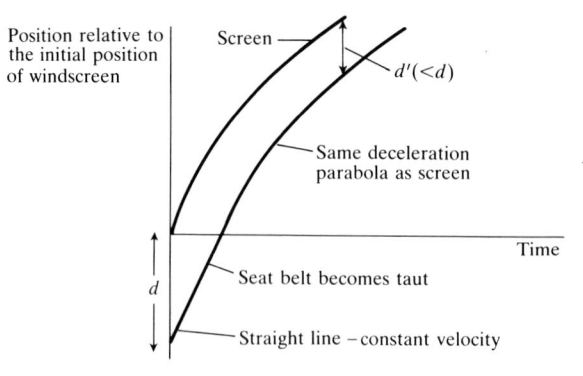

Fig. 3.22

* It should be noted that the same analysis applies to the seat-belt and the passenger (until the belt becomes taut) as to the windscreen and the unrestrained passenger. This means that the further the passenger travels before the belt tightens, the faster they will hit the belt and the greater its impact on them. Hence the generally held opinion that it is more dangerous to wear a seat-belt that is too slack than not to wear one at all.

3.5. Braking a car

- **Motion with constant acceleration $F = ma$**

How quickly can a car stop in an emergency? There are many factors which affect this.

The driver	—his reaction time
	—his skill
The car	—its speed
	—its weight
The brakes	—their type (disc, drum, etc.)
	—their quality
	—their state of maintenance
The tyres	—their type
	—their state of wear
The road	—the type of surface
	—weather conditions at the time
	—weather conditions previously (e.g. in long hot spells roads acquire a film of rubber and oil that becomes greasy in the first rain).
The nature of the emergency	—how quickly it becomes apparent.

- **Solution**

The total distance travelled between the moment when the driver perceives the emergency and the time when the car stops consists of two parts:

Thinking distance, i.e. the distance travelled before he actually starts breaking.
Braking distance, i.e. the distance the car then travels before coming to rest.

Thus we can write total distance = thinking distance + braking distance.

Mechanics

THINKING DISTANCE

If we assume that the car is travelling with a constant velocity v and if the thinking distance is d_t, then d_t depends solely upon v and the driver's reaction time, r seconds.

At constant velocity the distance travelled equals velocity times time, i.e.

$$d_t = vr.$$

Now r depends on many factors, the driver's age, state of health, state of mind, etc., but we could obtain an average value by experiment; this has been done and gives a value of about 0.68 s.

Some typical thinking distances are given in Table 3.2. Generally if v is given in m.p.h. then $d_t = 0.996 v$ (in feet) and $d_t = 0.304 v$ (in metres)—in other words, about 1 ft or 0.3 m for every mile per hour.

Table 3.2

Speed (m.p.h.)	Thinking distance	
	(ft)	(m)
30	30	9.1
45	45	13.7
60	60	18.2

BRAKING DISTANCE

First we need to know how the *effective* braking force, F, varies with the speed of the car. This depends on the type of brakes, discs or drums, and the relationship is shown approximately in Fig. 3.23(a) and (b).

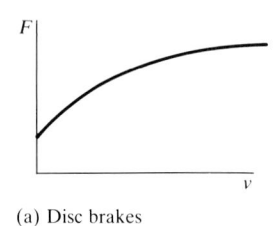

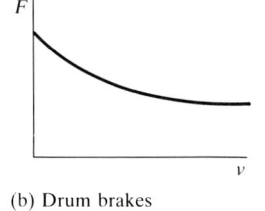

(a) Disc brakes   (b) Drum brakes

Fig. 3.23

This explains why most cars are fitted with a mixture of discs and drums, giving a combined effect as shown in Fig. 3.24. Although the line shown in Fig. 3.24 is not quite a horizontal straight line, it is not a bad approximation to *assume* that the effective braking force, F, is constant over the range of speeds we need to consider.

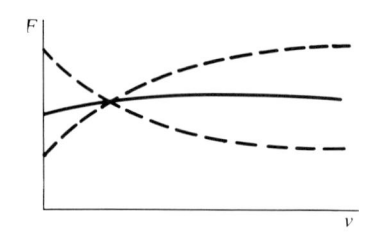

Fig. 3.24

What is this constant value? This obviously depends on the design of the car and the state of wear of the brakes, but for most cars it is about $\frac{2}{3}$ of the weight of the car.

If the mass of car is m kg, then $F = \frac{2}{3} mg$. So, using $F = ma$, where a is the car's acceleration during braking, then

$$-\tfrac{2}{3} mg = ma$$

where g is the acceleration due to gravity. (Note the $-$ sign, F is a braking force.)

Hence $v\,dv/dx = -\frac{2}{3}g$ where x is the distance travelled from the start of braking.

Integrating and setting $v = 0$ when $x = d_b$, the braking distance, we get

$$d_b = \frac{3v^2}{4g}.$$

Table 3.3 gives the braking distance for several speeds.

Table 3.3

Speed (m.p.h.)	Braking distance (ft)	(m)
30	44.2	13.61
45	99.5	30.61
60	176.9	54.42

Thus, approximately, we can write this law as

$$d_b = \frac{1}{20}v^2$$

where d_b is the braking distance in ft and v is the speed of the car in m.p.h.

Summarizing, the graphs of thinking and braking distances are shown in Fig. 3.25(a) and (b) and the total stopping distance is shown in Fig. 3.26.

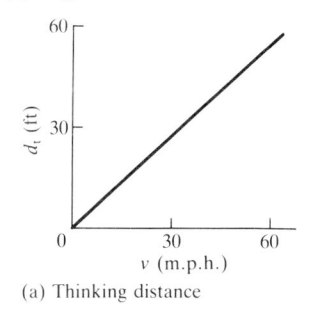

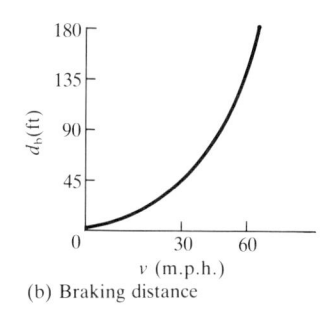

(a) Thinking distance

(b) Braking distance

Fig. 3.25

Stopping distance

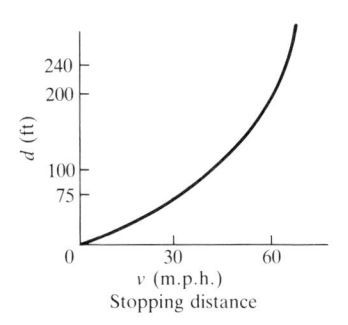

Fig. 3.26

Mechanics

Notes

It will be found that the values for thinking and braking distances given in the *Highway code* are those generated by this model.

The values chosen for r, the average reaction time, and the assumption that $F = \frac{2}{3}mg$ are only rough approximations that keep the later results as simple as possible.

• Exercises

1. Use the model to calculate the stopping distances for 10, 20, 30, ... 70 m.p.h. and compare them with the *Highway code*.

2. What would be the formula connecting d_b and v if d_b is in metres and v is in km/h?

3. How does the braking distance vary with the effective braking force? Suppose $F = pmg$ (i.e. the braking force is a fraction p of the weight of the car).

    (a) Find the braking distance from 30 m.p.h. for values of p between 0 and 2 at intervals of 0.2.

    (b) What is the formula connecting d_b with p if the initial speed is 30 m.p.h.?

    (c) What is the formula connecting d_b with p and v?

4. Suppose that, for a particular car, F is not constant but depends linearly on v, the relationship being approximated by the formula $F = (0.660 - 0.001\,v)mg$. Using the $v(dv/dx)$ integration method, find the new braking distances from 30 and 60 m.p.h.

3.6. Experiments in mechanics

In this section we outline three experiments that can be used to illustrate various phenomena that can be modelled using the theory of mechanics.

3.6.1. FALLING FREELY

Problem

Drop a large steel ball-bearing from a height of several metres above the floor. Find

(a) A simple relationship between time and distance fallen;

(b) How the speed changes as it falls.

Method

'Freeze' the motion by taking a series of photographs (all on one frame) by the light of a stroboscope flashing at regular interval. Put a 1.5-metre rule beside the falling ball so that you can measure where it is at each flash. From your measurements investigate the problem.

Teacher's notes

Apparatus

For the flash, use either a xenon stroboscopic lamp *or* photoflood lamps which illuminate the ball all the time together with a motor-driven strobe disc in front of the camera (slots cut symmetrically admit light at regular intervals). Take photographs with various frequencies if possible—from 10 to 50 exposures per second. With either lamp the room needs to be dark and the camera shutter remains open all the time.

The camera may be either a Polaroid model (great advantage is simplicity and speed) or a 35-mm single-lens reflex (modern development techniques produce a negative in less than 10 minutes, and the projected negative can be measured easily and accurately).

Technique

The distance of fall cannot be made much more than one metre. The ball should be dropped from the zero of the ruler. Before the ball has been falling for about 0.1 s (depending on its radius and the strobe frequency) the images will overlap.

Analysis

Two approaches to the solution suggest themselves

1. Calculate the average speed for each time interval, and then look at the average acceleration between one interval and the next. The figures will suggest constant acceleration (around 9.7 m s^{-1} is not difficult to obtain).

This means that the velocity/time graph of the motion will be a straight line, $v = 9.8t$, and the distance function can easily be found by considering area under this graph (i.e. integrating), giving $h = 4.9t^2$.

2. (a) We look for a simple polynomial of the form

$$h = k(t + a)^n$$

Mechanics

where a is the time which elapses between the ball being released and the instant of the first flash, which we make our zero for the time scale at this stage.

Examination of the table of experimental values rules out $n = 1$; a graph of $h^{1/2}$ against t gives a reasonable straight line, showing n is in fact 2, and giving k from the gradient and a from the intercept.

Shifting the zero of the time scale back to the instant of release then produces something like

$$h = 4.9t^2.$$

(b) The data from a good photograph with strobe frequency 50 cycles per second can be used for a numerical investigation of the limiting values of average speeds at various times in the fall. Alternatively, values calculated from the quadratic model $h = 4.9t^2$ can be used in the same way.

For example

$t(s)$	0.40	0.42	0.44	0.46	0.48	0.50
$h(m)$	0.784	0.864	0.949	1.037	1.129	1.225

gives average speed

Time interval (s)	Speed (m s^{-1})
0.4–0.5	4.41
0.4–0.48	4.31
0.4–0.46	4.22
0.4–0.44	4.13
0.4–0.42	4.00

suggesting that the speed at $t = 0.4$ s is 4 m s^{-1}. Similar calculations suggest a speed of 3 m s^{-1} at $t = 0.3$ s and 2 m s^{-1} at $t = 0.2$ s, and the relation $v = 10t$ or $v = 9.8t$.

3.6.2. THROWING SURPRISES

Problem

Take two cricket balls: drop one and at the same time throw the other horizontally. Which will hit the floor first?

Method

Adapt the experiment in §3.6.1 to take multiflash photographs of two ball-bearings moving at once: one can be propelled horizontally using a spring as shown in Fig. 3.27; the other dropped as before.

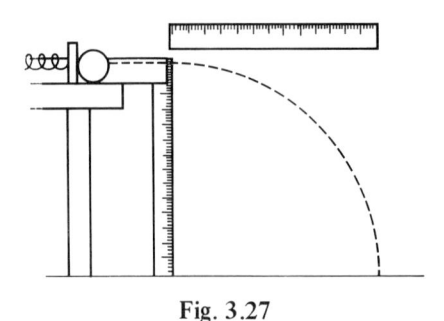

Fig. 3.27

Measure the vertical displacements of the dropped ball from the initial position for each flash. Measure the vertical and horizontal displacements for the projected ball. Compare the two. Describe the relations between the time elapsed and (a) the position of the balls; (b) their velocities.

Teacher's notes

Apparatus and technique

Suppliers of science apparatus (e.g. Griffin and George, Phillip Harris) manufacture a special device to project and drop two balls separately; alternatively it is easy enough to invent your own. The projected ball should travel at between 1 and 4 m s^{-1}.

Analysis

It will be strikingly clear from the photograph that the *vertical* motion of the two balls is the same. Measurements will show that the *horizontal* component of the velocity is constant. This suggests a vector model for the position something like

$$\mathbf{r} = 2t\mathbf{i} - 4.9t^2\mathbf{i}.$$

Extension

Defining average velocity in the time interval $t_2 - t_1$ as $(\mathbf{r}_2 - \mathbf{r}_1)/(t_2 - t_1)$, the velocity at an instant for the projected ball can be investigated numerically by looking for the limit of a sequence of averages over shorter and shorter intervals leading to the idea of acceleration.

$$\mathbf{r} = \begin{bmatrix} 2t \\ 4.9t^2 \end{bmatrix} \Rightarrow \mathbf{v} = \begin{bmatrix} 2 \\ 9.8t \end{bmatrix}$$

and thence to

$$\mathbf{a} = \begin{bmatrix} 0 \\ 9.8 \end{bmatrix}.$$

3.6.3 THE SPEED OF AN OSCILLATOR

Problem

Find a model for the displacement and speed of a body oscillating between two springs by direct measurements of its position.

Method

Make measurements of the position of a dynamics trolley during one-half of an oscillation by attaching the tape from a ticker tape vibrator to it.

Draw a graph of displacement against time for the half-oscillation; extrapolate your graph to show what you expect for the next two half-oscillations. Now shift the zero of the time scale to the *end* of the first half-oscillation (i.e. draw a new displacement axis and re-number the time axis). Likewise move the time axis up so that the zero of the displacement axis falls at the central, equilibrium position. You should recognize the shape of the graph. Write down its equation as accurately as possible by considering: (a) how often it repeats; (b) the largest and smallest values of the displacement about the centre.

Draw tangents to your graph at suitable intervals to enable you to draw a new graph of velocity against time (the velocity will be the gradient of the tangent). Apply similar reasoning to find the equation for this graph.

Mechanics

Teacher's notes

Apparatus and technique

The surface of the bench should be compensated for friction in one direction (that of the first half-oscillation) by tilting it a little until the trolley, without springs, rolls at constant speed down it once it is given a push. A typical arrangement (see Fig. 3.28) will have the stands about 1 m apart and the trolley tethered to two or three pairs of linked springs; the initial amplitude of the oscillation should be arranged to be as near 10 cm as possible.

The ticker tape is fastened to the trolley with sellotape and the vibrator should be started just before the trolley released.

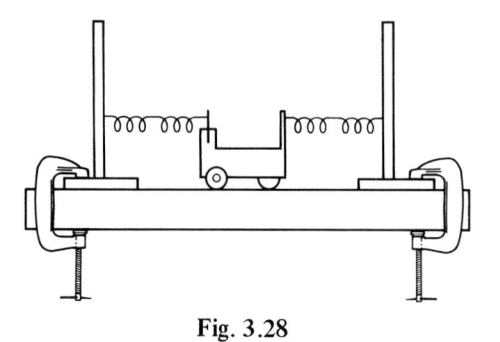

Fig. 3.28

Analysis

This experiment is best done before a student has experience of Newton's laws and simple harmonic motion.

The experimental graph, relative to the new axes, will look something like Fig. 3.29(a). Comparison with $x = \cos t$ (Fig. 3.29(b)) shows that two one-way stretches have taken place. Investigation of one-way stretches of $x = \cos t$ soon leads to the conclusions that a scale-factor of A vertically produces $x = A \cos t$, and a scale factor of $i/12$ horizontally produces $x = \cos kt$. Applied together, we deduce a model of $x = 10 \cos(180t)$ for the oscillator's displacement.

The velocity graph is of the form $v = -B \sin kt$ and, if the student has not met radian measure, this may well be a good opportunity to introduce it, ending up with neat results.

(a) Experimental graph   (b) $x = \cos t$

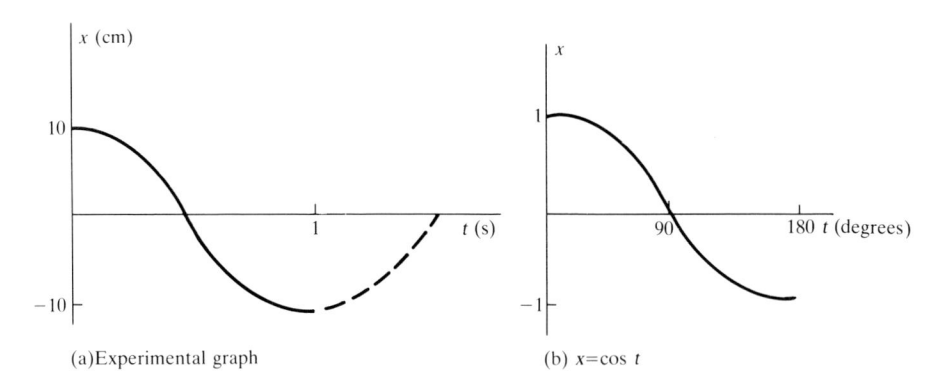

Fig. 3.29

3.7. Kepler's law

• **Newton's Second Law; empirical modelling**

Table 3.4 gives details of the average distance from the sun, R, and period of revolution around the sun, T, of the known planets.

Table 3.4

Planet	Distance from the sun R (10^6 km)	Period of revolution around the sun, T (days)
Mercury	57.9	88
Venus	108.2	225
Earth	149.6	365
Mars	227.9	687
Jupiter	778.3	4 329
Saturn	1427	10 753
Uranus	2870	30 660
Neptune	4497	60 150
Pluto	5907	90 670

In 1601 the German astronomer and science-fiction writer, Johann Kepler, became the Director of the Prague Observatory and, after studying the relative motion of the planet Mars, by 1609 he had formulated his first two laws:

1. Each planet moves on an ellipse with the sun at one focus;
2. For each planet, the line from the sun to the planet sweeps out equal areas in equal times.

Kepler spent the next decade in verifying these two laws for other planets and, in 1619, he published his third law, dedicated to James I of England:

3. The squares of the orbital periods of the planets vary at the cubes of their mean distance from the sun.

Show that Kepler's third law provides a good model for the data (a) by fitting the data (like Kepler), and (b) by applying Newton's laws of motion.

• **Solution**

An empirical model

A plot of the period T against the distance R using Table 3.4 certainly indicates a non-linear law relating to these two variables (see Fig. 3.30).

If we assume a power law of the form

$$T = KR^\alpha$$

where K and α are constants, then we can easily check this with the data; for, taking logs,

$$\log T = \log(KR^\alpha) = \log K + \alpha \log R.$$

Thus a plot of $\log T$ against $\log R$ should be a straight line. The gradient and intercept of this

Mechanics

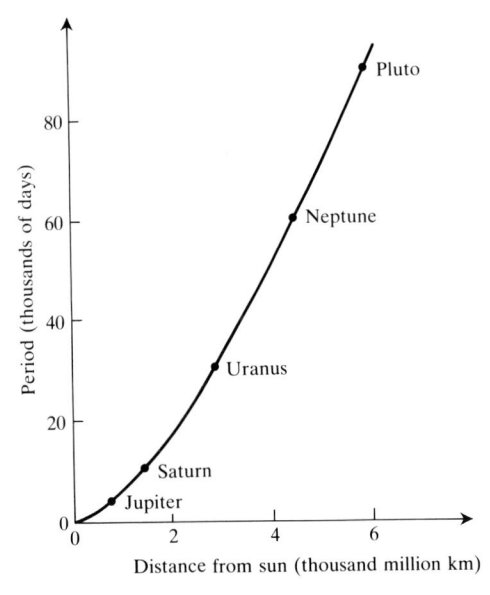

Fig. 3.30

line will give α and K. Figure 3.31 shows that the data points do lie on a straight line, and the gradient is approximately 1.5. Thus we conclude that

$$T = KR^{1.5} \qquad (3.17)$$

which agrees with Kepler's third law.

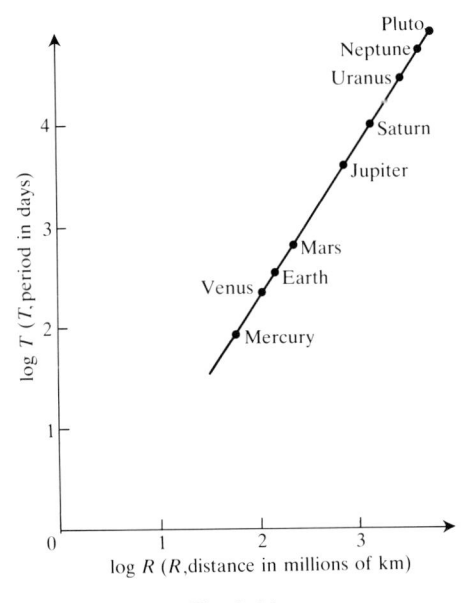

Fig. 3.31

A theoretical model

We now look at the theoretical model based on Newton's laws of motion and gravity. Assuming that a planet moves on a *circle* of radius R we have

$$m\ddot{\mathbf{r}} = -\frac{\gamma mM}{R^2}\mathbf{e}_r \qquad (3.18)$$

where m is the planet's mass, M the sun's mass, and γ the universal gravitational constant (see Fig. 3.32). The vectors $\mathbf{e}_r$ and $\mathbf{e}_\theta$ are unit vectors along and perpendicular to the radius vector.

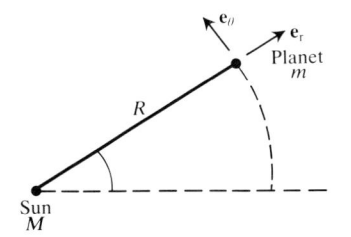

Fig. 3.32

Now
$$\dot{\mathbf{r}} = R\dot{\theta}\mathbf{e}_\theta$$
$$\ddot{\mathbf{r}} = R\ddot{\theta}\mathbf{e}_\theta - R\dot{\theta}^2\mathbf{e}_r$$

and so
$$R\ddot{\theta}\mathbf{e}_\theta - R\dot{\theta}^2\mathbf{e}_r = -\gamma\frac{M}{R^2}\mathbf{e}_r$$

Equating coefficients of $\mathbf{e}_\theta$ and $\mathbf{e}_r$ gives $\ddot{\theta} = 0$ so $\dot{\theta}$ is constant and
$$R\dot{\theta}^2 = \gamma M/R^2,$$

i.e.
$$\frac{d\theta}{dt} = (\gamma M)^{1/2}/R^{3/2}. \tag{3.19}$$

If we integrate (3.19) round a complete cycle, then θ goes from 0 to 2π whilst t goes from 0 to T, the periodic time; then we have

$$\int_0^{2\pi} d\theta = \int_0^T [(\gamma M)^{1/2}/R^{3/2}]\, dt$$

$$2\pi R^{3/2} = (\gamma M)^{1/2} T. \tag{3.20}$$

and so we have again derived Kepler's third law.

• Exercises

1. In 1766, Titius gave the formula
    $$R = (4 + 3 \times 2^n)/10$$
    for the distance of planet number n from the Sun ($n = -\infty$ for Mercury, $n = 0$ for Venus, $n = 1$ for Earth, etc. missing $n = 3$) in units such that $R = 1$ for the Earth.

    Is this formula an accurate one? (It is known as Bode's Law.)

2. If planetary motion is governed by an inverse *cube* law is Kepler's third law changed?

Mechanics

3.8. Industrial location

• Centre of mass

Most food shops are now part of a chain of stores, and one of the problems faced by a chain of stores is where to place the warehouse which serves all the stores in the area. Costs must be minimized, and the main contribution to these costs is related to the transportation of the goods to and from the shops.

These transportation costs will clearly depend on

1. Distance travelled;
2. Volume of goods transported.

Formulate a model for these costs, and analyse the resulting model.

• Solution

A typical problem is to determine the optimum location for a single depot for a given demand in the shops in the region. Suppose the depot is located at (x, y) and that this depot supplies n shops positioned at $i = 1, 2, \ldots n$. The total costs, T.C., are derived from

$$\text{T.C.} = \sum_{i=1}^{n} C_i$$

where C_i is the cost of transport of goods (per unit weight) from the depot to shop i. We assume that

$$C_i = \alpha w_i d_i$$

where α is a constant, w_i is the demand (in unit time) by shop i, and d_i is the distance from the depot to shop i. Hence

$$d_i = [(x - x_i)^2 + (y - y_i)^2]^{1/2}$$

and our problem is to find the optimum location (x, y) of the depot which minimizes

Mechanics

$$\text{T.C.} = \alpha \sum_{i=1}^{n} w_i [(x - x_i)^2 + (y - y_i)^2]^{1/2}$$

This is a function of two variables, x and y, and it does not possess analytic solutions in general. We can nevertheless make some progress, but before returning to the general problem we look first at some special cases.

The case $n = 2$ will have the depot location along the straight line joining the two shops. The exact location will depend on the relative shop demands or weights (see below). For $n = 3$, if $w_1 = w_2 = w_3$, the solution is relatively straightforward. Provided the triangle formed by the three shops has all its angles less than 120° the depot will be located such that the lines to the shops all make angles of 120°. If the triangle has an angle greater than 120° then the depot will be located at that position. These solutions are illustrated in Fig. 3.33.

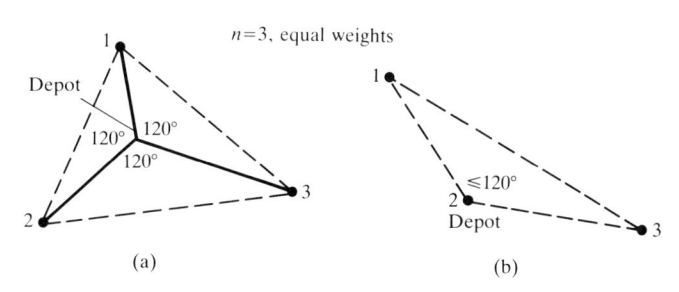

Fig. 3.33

Similarly for equal weights for the case $n = 4$, the depot will be located at the intersection of the diagonals provided they do in fact intersect. If not, the quadrilateral formed by the shops position is non-convex, and the depot will be located at the 'internal' shop as illustrated in Fig. 3.34.

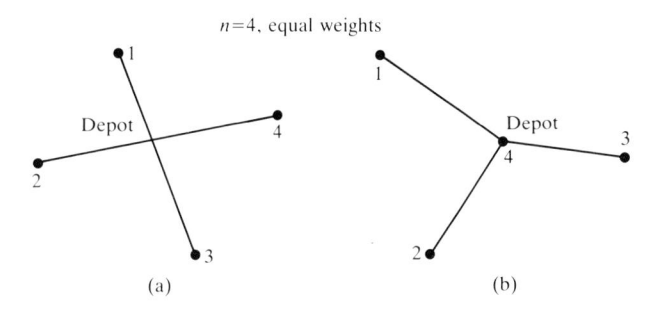

Fig. 3.34

One suggestion for the solution of the general problem was to locate the depot at the centre-of-gravity of the shops' 'weights', i.e.

$$\bar{x} = \frac{\sum_{i=1}^{n} w_i x_i}{\sum_{i=1}^{n} w_i}, \quad \bar{y} = \frac{\sum_{i=1}^{n} w_i y_i}{\sum_{i=1}^{n} w_i} \tag{3.22}$$

but, although this position is often close to the true optimal solution, it is not the solution. We can clearly see this for the costs of two customers, illustrated in Fig. 3.35. Here

$$\text{T.C.} = w_1 x + w_2 (D - x)$$
$$= (w_1 - w_2) x + w_2 D$$

Mechanics

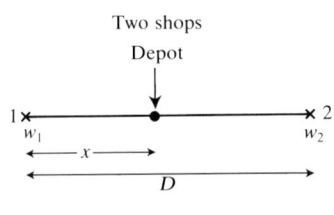

Fig. 3.35

and, for $0 \leqslant x \leqslant D$, the minimum value of T.C. occurs

1. at $x = 0$, if $w_1 > w_2$;
2. at $x = D$, if $w_2 > w_1$;
3. any value of x, $0 \leqslant x \leqslant D$, if $w_1 = w_2$.

So clearly the centre-of-gravity solution (3.22) is not correct.

There is, however, a mechanical analogue which does give the correct optional solution. It is illustrated in Fig. 3.36. A map of the region is pasted onto hardboard and holes made at each of the shop positions. Strings are passed through each of the holes and weights proportional to w_i, attached to shop i whilst the other ends of the string are attached to a ring. If the length of the string through shop i is l_i, the potential energy of the system

$$V = \sum_{i=1}^{n} w_i(l_i - d_i) = \sum_{i=1}^{n} w_i l_i - (\text{T.C.})/\alpha$$

using (3.21). Now minimum potential energy position will correspond to minimum T.C. so that if the system is released from rest it will take up its equilibrium position with minimum potential energy, and so we have the desired optimal location for the depot.

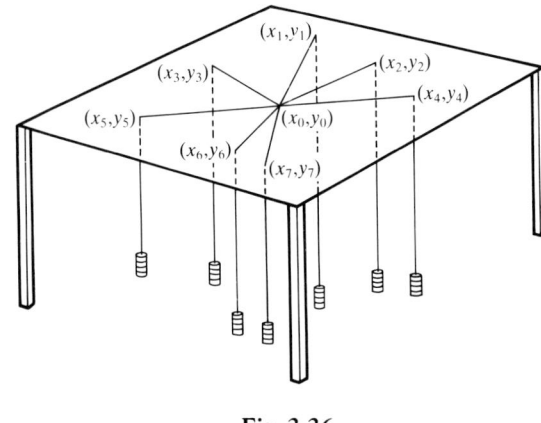

Fig. 3.36

The method is crude and simple, but it is effective and it is comparatively easy to see the effect of changes to the system.

• Exercises

1. Apply the analysis to a chain of shops in your area. Construct the mechanical analogue solution, and consider the validity of your solution.

2. Verify the $n = 3$, $w_i = w_2 = w_3$ solution stated in the text.

4. Probability and statistics

Probability and statistics

4.1. Are you being served

- **Binomial distribution; normal approximation to binomial**

If a coin is tossed 100 times and the result is 46 heads and 54 tails, it would generally be accepted as a typical result of a random experiment (and indeed deviations of this order would be expected about 40 per cent of the time).

But what if the result were 40 heads and 60 tails? Should we now suspect the coin to be biased?

The example which follows concerns the fairness of a roulette wheel, which has 37 equally-spaced numbers, 18 of which are black, 18 red, and one (zero) green. In 100 spins of the wheel, a black number is obtained on 63 occasions. Does this suggest that the wheel is biased in favour of the black numbers?

- **Solution**

Consider the null hypothesis that the wheel is unbiased, with the probability of a black number occurring being $p = 18/37$, the alternative hypothesis being that $p > 18/37$.

The number of occurrences of a black number will be binomially distributed with

$$n = 100, \quad p = \frac{18}{37}, \quad q = \frac{19}{37}.$$

Since $p > 0.1$ and $np > 5$, the *normal* approximation to the *binomial* can be used with

$$\mu = np = 100 \times \frac{18}{37}$$

and

$$\sigma = \sqrt{(npq)} = \sqrt{\left(100 \times \frac{18}{37} \times \frac{19}{37}\right)}.$$

If x is the number of times a black number occurs in 100 spins,

$$z = \frac{x - \mu}{\sigma}$$

will be the standardized normal variate (mean = 0, SD = 1). In this case

$$z = \frac{63 - 100 \times (18/37)}{\left(100 \times \frac{18}{37} \times \frac{19}{37}\right)^{\frac{1}{2}}}$$

$$\simeq 2.87.$$

Referring to normal distribution tables, it will be seen that under the null hypothesis the chance of such a result is about

$$(1 - 0.99795) \times 100 \text{ per cent} \simeq 0.2 \text{ per cent}.$$

This is so small that, unless there is a good reason for doing otherwise, we must reject the null hypothesis and conclude that the wheel favours the black numbers.

Probability and statistics

- **Exercises**

1. In 200 tosses of a coin, head is obtained 80 times. Is this evidence that the coin is biased?

   Note: It must be decided in advance whether to test for fairness of the coin (two-tailed test) or to test whether the coin is biased in favour of tails (one-tailed test).

2. How many times must a fair coin be tossed to ensure that there is a 95 per cent chance of obtaining a head?

   Hint: Computation is simplified by using the relation
   $$P \text{ (at least 1 head)} = 1 - P \text{ (no heads)}.$$

3. In a game of snakes and ladders, a player can start only after throwing a six. How many throws are needed to give a 75 per cent chance of starting the game

4. A travel firm operates 20 coaches which break down randomly. If each coach is out of service 10 per cent of the time as a result of breakdowns, for what percentage of the time will

   (i) All coaches be in service;
   (ii) Three coaches be out of service;
   (iii) More than three coaches be out of service.

- **Further reading**

Murdoch, I. and Barnes, J. A. (1973). *Statistics: problems and solutions*, Macmillan, Basingstoke.

4.2. Quality control

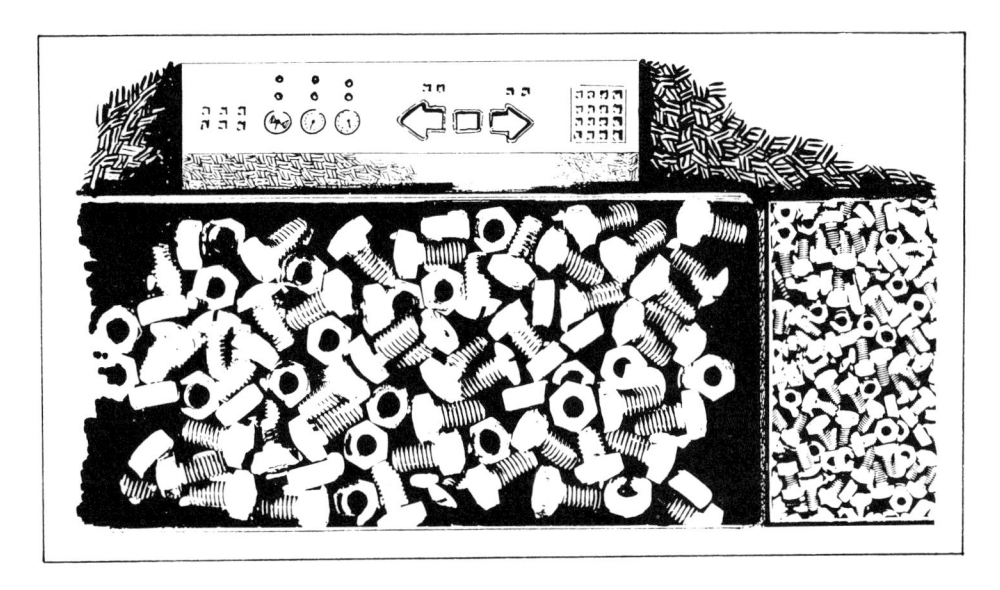

• **Normal distribution; distribution of sum of normal variates**

Machinery may be set to produce articles of a specified size, but in reality there is a certain amount of variation. Thus a process may produce bolts with a mean length of 25 mm and a standard deviation of 0.025 mm.

This variation has considerable implications in the manufacturing industry. For example, suppose washers are to be made to fit a bolt of diameter 20 mm, and the machine is set to produce washers with a mean diameter of 21 mm and a standard deviation of 1 mm. Then about 16 per cent of washers produced would be too small to fit the bolt; the manufacturer would need to reduce the variability or else set the machine at a higher mean value.

The following problem illustrates the ideas.

A sugar manufacturer markets his products in 2 kg bags. The packing process is normally distributed with a standard deviation of 0.04 kg.

1. If trade regulations specify that no more than 3 per cent of the bags may be 5 per cent under the nominal weight of 2 kg, what should the process average be set at?

2. The manufacturer decides to offer the product to bulk buyers in jumbo-packs containing 16 2-kg bags. What percentage saving can be made if the jumbo-packs are subject to the same trade regulations as in (1).

• **Solution**

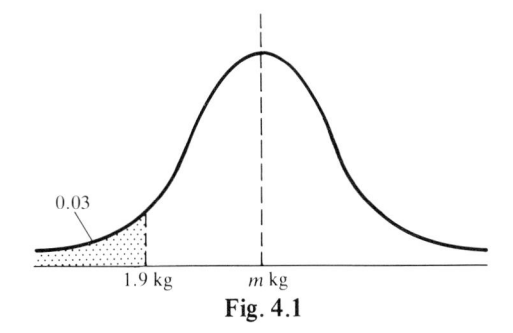

Fig. 4.1

Probability and statistics

1. Suppose the machine setting is m kg. No more than 3 per cent of the bags must be less than 95 per cent of 2 kg, i.e. 1.9 kg (see Fig. 4.1). From tables, a 3 per cent tails is cut off at a standardized value of -1.88. Hence,

$$\frac{1.9-m}{0.04} = -1.88 \text{ kg}$$

so

$$m = 1.9 + 0.04\,(1.88)$$
$$\simeq 1.975 \text{ kg}.$$

The machine should be set at 1.975 kg.

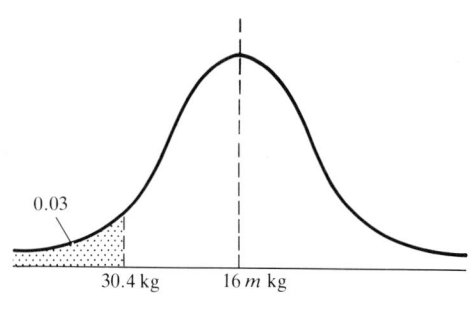

Fig. 4.2

2. Mean of jumbo-pack $= 16 \times 2$ kg $= 32$ kg. SD of jumbo-pack $= \sqrt{16} \times 0.04$ kg $= 0.16$ kg. If M kg is the setting now required, and no more than 3 per cent of the jumbo-packs must have a weight below 95 per cent of 32 kg, i.e. 30.4 kg, (see Fig. 4.2) then

$$\frac{30.4 - 16M}{0.16} = -1.88 \text{ kg}$$

so

$$M = \frac{30.4 + 0.16(1.88)}{16}$$
$$\simeq 1.919 \text{ kg}.$$

The percentage saving

$$= \frac{(1.975 - 1.919)}{1.975} \times 100 \text{ per cent}$$
$$\simeq 2.8 \text{ per cent}$$

• Exercises

1. In a machine fitting caps to bottles, the force (torque) applied is distributed normally with mean 8 units and standard deviation 1.2 units. The breaking strength of the caps has a normal distribution with mean 12 units and standard deviation 1.6 units. What percentage of caps are likely to break on being fitted?

2. The maximum payload of a light aircraft is 350 kg. If the weight of an adult is normally distributed with a mean 75 kg and SD 15 kg, and the weight of a child is normally distributed with mean 23 kg and SD 7 kg, what is the probability that the plane can take off safely with

(a) four adult passengers;

(b) four adult passengers and one child?

In each case, what is the probability that the plane can take off if 40 kg of baggage is carried (in total)?

3. Two types of metal rods are produced by machines, one having a length of 50 mm ± 0.2 mm, the other having a length of 20 mm ± 0.2 mm. These are 95 per cent tolerances.

In an assembly process one of each type is selected at random, and the two rods placed end-to-end. If the distributions of the two types are normal, find the 95 per cent tolerances for the combined length.

4. A salesman has to make 15 calls a day. Including time for travelling, he spends on average 30 minutes with each customer, this time having a SD of 6 minutes.

(i) If his working day is 8 hours long, what is the probability that on a given day he will have to work overtime?

(ii) Find 99 per cent confidence limits for the limits of his 'free' time in a 5-day week.

• **Further reading**

Murdoch, J. and Barnes, J. A. (1973). *Statistics: problems and solutions*, Macmillan, Basingstoke.

4.3. Blood donors

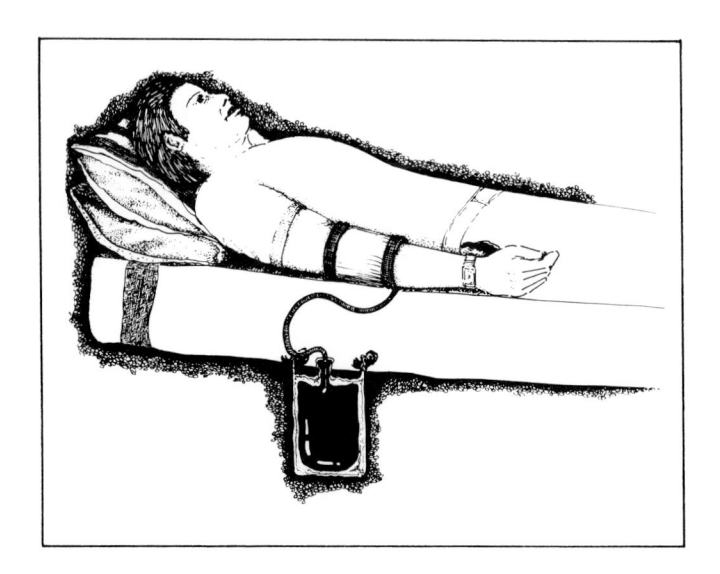

• **Probability**

Before an individual can receive a blood transfusion, the blood of both donor and recipient must be matched according to certain criteria. There are substances called *antigens* which may or may not be present in the blood of an individual. Blood types are classified by the presence or absence of three of these antigens, called A, B, and Rh, as indicated in Table 4.1 (which also gives the frequency of occurrence of each type in the United States in 1970).

Table 4.1. Table of blood classifications by the antigens A, B, Rh, including the relative occurrence of each type in the USA in 1970

Blood types	Antigens present	Frequency (per cent)
O−	—	6.6
A−	A	6.3
B−	B	1.5
AB−	A, B	0.6
O+	Rh	37.4
A+	A, Rh	35.7
B+	B, Rh	8.5
AB+	A, B, Rh	3.4

A person cannot receive blood of a type containing an antigen which is not present in his own blood. For example,

1. Types A+ and B+ are incompatible in either direction;
2. A person of type B+ can receive blood of type B−, but not vice versa.

What is the probability that a person selected at random can donate blood to someone with a given blood type (using 1970 US data in Table 4.1)?

Probability and statistics

• Solution

The data of Table 4.1 can be illustrated on a Venn diagram (Fig. 4.3). Consider the problem of determining the probability that a person selected at random can donate blood to someone of type B+.

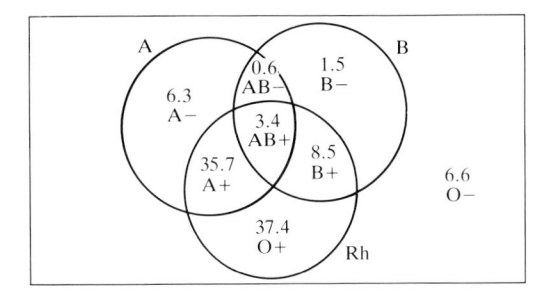

Fig. 4.3. Venn diagram showing distribution (per cent) of blood types in US, 1970.

Since type B+ contains antigens B and Rh, the only suitable donors are those with types O−, O+, B−, and B+ (since the remaining types contain antigen A which is absent from type B+). Figure 4.3 shows that the approximate probability is $(6.6 + 37.4 + 1.5 + 8.5)/100 = 0.54$. In a similar way, the probabilities can be determined for each blood type, giving the results shown in Table 4.2.

Table 4.2. Probability of random selection producing suitable donors for a given blood-type

Blood type	Probability of occurrence	Suitable donor types	Probability that random selection produces suitable donor
O−	0.066	O−	0.066
A−	0.063	O−, A−	0.129
B−	0.015	O−, B−	0.081
AB−	0.006	O−, A−, B−	0.144
O+	0.374	O−, O+	0.440
A+	0.357	O−, O+, A−, A+	0.860
B+	0.085	O−, O+, B−, B+	0.540
AB+	0.034	O−, O+, A−, A+, B−, B+	1.000

Inspection of Table 4.2 reveals some interesting results.

1. Although types O+ and A+ occur with almost identical frequency (37 per cent and 36 per cent, respectively), the probability of finding suitable donors for type A+ is almost twice that for type O+ (0.86 as against 0.44):
2. Although type AB+ occurs with a frequency of only 3.4 per cent, there is no difficulty finding suitable donors!

• Exercises

1. Calculate the probability that an individual selected at random is a suitable recipient for blood type A+ (that is, can receive blood of type A+).
2. Repeat 1 for each of the other seven blood types, producing a table similar to Table 4.2.
3. Comment on the results obtained in 2.

4.4. Intelligence quotients (IQ)

• **Normal distribution**

IQ tests are designed so that the mean is 100, the standard deviation is 15, and the distribution is normal.

1. What proportion can be expected to have IQs
   (a) greater than 120;
   (b) less than 90;
   (c) between 70 and 130?
2. What IQ will be exceeded by
   (a) 1 per cent of the population;
   (b) 0.1 per cent of the population;
   (c) 90 per cent of the population?
3. What percentage of the population can be expected to have IQs within three standard deviations of the mean?

• **Solution**

(In the following analysis, x denotes a standardized normal variate, and $A(x)$ the area under the normal probability curve to the left of x.)

1. (a) $$x = \frac{120 - 100}{15} \simeq 1.33$$

From tables, $A(x) \simeq 0.9082$. Hence the percentage of population having an IQ exceeding 120 is $100(1 - 0.9082) \simeq 90$ per cent.

(b) $$x = \frac{90 - 100}{15} \simeq -0.67.$$

$$A(x) = 0.2514.$$

Hence the percentage of population having an IQ less than 90 is $100(0.2514) \simeq 25$ per cent.

(c) $$x_2 = \frac{130 - 100}{15} = 2, \qquad x_1 = \frac{70 - 100}{15} = -2$$

$$A(x_2) = 0.97725, \qquad A(x_1) = 0.02275.$$

Hence percentage of population having an IQ between 70 and 130 is $100(0.97725 - 0.02275) \simeq$ $\simeq 95$ per cent.

2. (a) $$A(x) = 0.99$$

so

$$x \simeq 2.33$$

Hence required IQ $= 100 + 15(2.33) \simeq 135$.

(b) $$A(x) = 0.999$$

Solutions to the exercises

so
$$x \simeq 3.1.$$
Hence required IQ $= 100 + 15(3.1) = 146.5$

(c) $$A(x) = 0.1$$
so
$$x \simeq -1.28.$$
Hence required IQ $= 100 - 15(1.28) \simeq 81$.

3. $$x_1 = -3, \qquad x_2 = 3.$$
$$A(x_1) = 0.00135, \quad A(x_2) = 0.99865.$$
Hence, required percentage $= 100(0.99865 - 0.00135) \simeq 99.7$ per cent.

- **Further reading**

Murdoch, J. and Barnes, J. A. (1973). *Statistics: problems and solutions*, Macmillan, Basingstoke.

Solutions to the exercises

Solutions to the exercises

1. Functions

1.1. OSCILLATIONS IN NATURE

1. The amplitude for the oscillations is 1 and the period is 5 seconds. We could model the pressure, p/p_0, in the lungs at time t by

$$\frac{p}{p_0} = \sin\left(\frac{2\pi}{5}t - \pi\right).$$

A graph of this function is shown in Fig. 5.1.

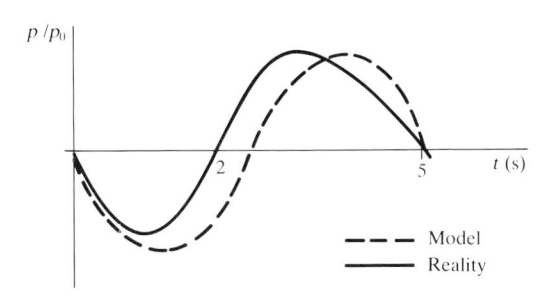

Fig. 5.1

1.2. PARTY REPRESENTATION AND ELECTION RESULTS

1. Rearrangement of eqn (1.3) gives the form

$$y = \frac{x^3}{3x^2 - 3x + 1}.$$

The graph of the cube law is shown in Fig. 5.2. For values of x between 0.4 and 0.6 the graph in Fig. 5.2 can be seen to be almost linear. In a two-party system it is common for each party to poll between 40 and 60 per cent of the votes cast.

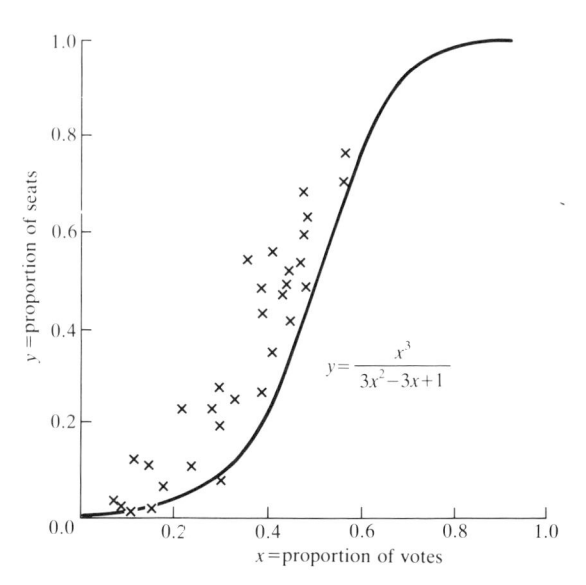

Fig. 5.2

Solutions to the exercises

1.3 GROWTH RATES

1. (i) The following table shows the number of ewes for the first few years.

t (years)	Number of ewes ($P(t)$)
0	100
1	160
2	256
3	410

In symbolic form we have

$$\text{Number of ewes at end of year } t + 1 = P(t) + \tfrac{1}{2}P(t) \times 1.2$$
$$= 1.6 P(t)$$

Thus

$$P(1) = 1.6 P(0),$$
$$P(2) = 1.6 P(1) = 1.6^2 P(0),$$
$$P(3) = 1.6 P(2) = 1.6^3 P(0),$$

or, in general,

$$P(t) = P(0) 1.6^t = 100(1.6)^t.$$

(ii) In terms of $\log s$ (to base 10) we have

$$\log P = \log(100) + t \log(1.6)$$
$$= 2 + 0.2t.$$

(iii) P exceeds 600 when

$$\log P > \log(600)$$

or

$$2 + 0.2t > \log(600)$$

so

$$t > [\log(600) - 2]/0.2 = 3.89,$$

i.e the population exceeds 600 after 4 years.

3. Figure 5.3 shows a graph of $\log P$ plotted against t. The population in the year 2000 will be about 530-560 million people.

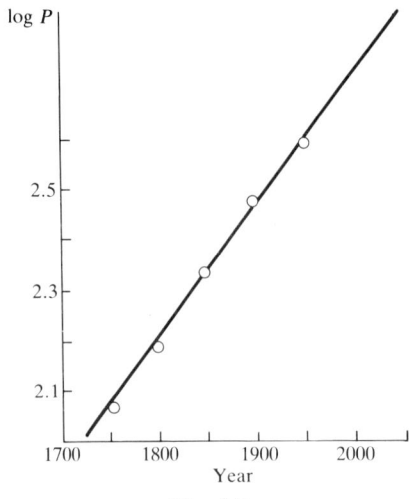

Fig. 5.3

Solutions to the exercises

1.6 SHORTEST PATHS ACROSS RECTANGULAR CITIES

1. (i) To get from A to C we have to travel east for 2 blocks and south for 2 blocks. So we have to work out the number of ways of choosing 2Es and 2Ss to fill 4 boxes, i.e. the coefficient of x^2y^2 in $(x+y)^4$.

   (ii) Similarly there are $^3C_1(=^3C_2)$ ways of getting from C to B, i.e. three ways.

1.6 GRAVITATIONAL FORCE

1. The third-order approximation to the weight is

$$W_{third} = mg\left(1 - 2\left(\frac{\delta}{a}\right) + 3\left(\frac{\delta}{a}\right)^2 - 4\left(\frac{\delta}{a}\right)^3\right).$$

The third-order approximation differs from the second-order approximation by 10 per cent when

$$\left|\frac{W_{third} - W_{second}}{W_{second}}\right| \geqslant 0.1,$$

i.e. when

$$\frac{4\left(\frac{\delta}{a}\right)^3}{\left(1 - 2\left(\frac{\delta}{a}\right) + 3\left(\frac{\delta}{a}\right)^2\right)} \geqslant 0.1,$$

i.e.

$$4\left(\frac{\delta}{a}\right)^3 - 0.3\left(\frac{\delta}{a}\right)^2 + 0.2\left(\frac{\delta}{a}\right) - 0.1 \geqslant 0.$$

This inequality can be solved in many ways, e.g. sketch of graph, bisection method, Newton–Raphson, and so on. The solution for δ/a is 0.258 (to 3 decimal places) and, with $a = 4000$ miles, the value of δ is 1032 miles.

1.7 INVESTMENT AND BORROWING

1-3. In each of the exercises we use the formula

$$A_n = \left(P + \frac{p}{R}\right)(1+R)^n - \frac{p}{R}.$$

The values of P, p, R, and n and the answer to each exercise is shown in the following table.

Exercise	P	p	R	n	Answer (£)
1	100	100	0.05	20	3572
2	150	150	0.1	9	2390
3	200	200	0.08	14	5430

4. The loan would be 80 per cent of £20 000 = £16 000 ($=-P$), and $A_0 = 0$. Also $R = 0.14$ and $n = 25$. From the formula above,

$$p = \frac{RP(1+R)^n}{1-(1+R)^n} = £2328 \quad \text{(to nearest pound)}.$$

Hence the monthly repayment is £194 (to nearest pound).

5. The distance travelled by the bouncing ball is given by

$$d = h + 2[\tfrac{2}{5}h + \tfrac{2}{5}((\tfrac{2}{5}h) + \tfrac{2}{5}(\tfrac{2}{5}h)) + \ldots],$$

i.e.

Solutions to the exercises

$$d = h + 2[\tfrac{2}{5}h + (\tfrac{2}{5})^2 h + (\tfrac{2}{5})^3 h + (\tfrac{2}{5})^4 h + \ldots].$$

Now, $S = \tfrac{2}{5}h + (\tfrac{2}{5})^2 h + (\tfrac{2}{5})^3 h \ldots$ is the sum of a geometric series with first term $= a = \tfrac{2}{5}h$ and common ratio $= r = \tfrac{2}{5}$, but the number of terms is infinite. So $S = a/(1-r)$, provided $r < 1$. In this case $r = \tfrac{2}{5}$ and then

$$S = \frac{(\tfrac{2}{5}h)}{(\tfrac{3}{5}h)} = \frac{2h}{3}.$$

Hence the total distance travelled by the bouncing ball (Fig. 5.4) is

$$d = h + 2S = \frac{7h}{3}.$$

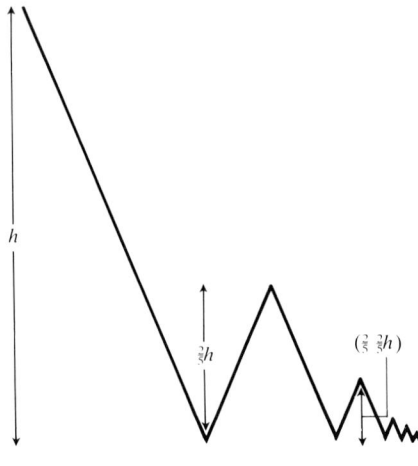

Fig. 5.4

1.8 SCREE SLOPES

1. If $d = 2$, $h_1 = 50$, and $h_2 = 30$, then

$$\tan \alpha = \frac{(h_1 - h_2)^2}{2h_1 d} = \frac{20^2}{20 \cdot 50 \cdot 2} = 2.$$

Hence $\alpha = 63.4°$.

2. Again equating areas we have

$$\text{Area of SBCT} = \text{Area QRPBC}.$$

The shape of QRPBC is made up of two triangles, QRW and RBP, and a rectangle CBRW.

Thus the area of QRPBC = area of RBP + area of QRW + area of CBRW

$$= \tfrac{1}{2} BP \cdot BR + \tfrac{1}{2} WR \cdot WQ + BC \cdot CW$$

Now

$$BC = d; \quad CQ = h_1 - h_2.$$

Hence

$$WQ = d \tan \alpha, \quad CW = h_1 - h_2 - d \tan \alpha$$

and

$$BP = \frac{BR}{\tan \beta} = \frac{h_1 - h_2 - d \tan \alpha}{\tan \beta}.$$

Thus the area of QRPBC

$$= \frac{(h_1 - h_2 - d \tan \alpha)^2}{2 \tan \beta} + \frac{d^2 \tan \alpha}{2} + d(h_1 - h_2 - d \tan \alpha)$$

Solutions to the exercises

$$= d^2\left(\frac{\tan^2\alpha}{2\tan\beta} - \frac{\tan\alpha}{2}\right) + d(h_1 - h_2)\left(1 - \frac{\tan\alpha}{\tan\beta}\right) + \frac{(h_1 - h_2)^2}{2\tan\beta}$$

and the area of SBCT $= dh_1$.

Equating these two areas we have the following quadratic equation for d,

$$\frac{\tan\alpha}{2}\left(\frac{\tan\alpha}{\tan\beta} - 1\right)d^2 - \left\{(h_1 - h_2)\frac{\tan\alpha}{\tan\beta} + h_2\right\}d + \frac{(h_1 - h_2)^2}{2\tan\beta} = 0.$$

Solutions to the exercises

2. Calculus

2.1 THE SHAPE OF A TIN CAN

Suppose that the base is a square of side length a and the height is h. Then the surface area A and volume V of the cuboid is given by

$$A = 2a^2 + 4ah$$
$$V = a^2 h.$$

Hence

$$A = 2a^2 + \frac{4V}{a}.$$

The minimum surface area is given when $(dA/da) = 0$. So

$$\frac{dA}{da} = 4a - \frac{4V}{a^2} = 0$$

and

$$a = V^{1/3}.$$

This is indeed a minimum since

$$\frac{d^2 A}{da^2} = 4 + \frac{8V}{a^3} = 12 > 0.$$

The minimum surface area is then $6a^2$.

2.2 FLEET SIZE FOR CAR LEASING COMPANY

1. For a lease charge of £1500 per car the formula for income becomes

$$I = 1500x, \qquad \text{if } 0 \leqslant x \leqslant 10$$

$$= \left(1500 - \frac{1}{100} 1500(x - 10)\right)x, \quad \text{if } x > 0.$$

Clearly $(dI/dx) = 0$ for $x = 55$ as before.

The formula for maximum profit becomes

$$P = 1500 - 500x, \qquad \text{if } 0 \leqslant x \leqslant 10$$
$$= 1650x - 15x^2 - 500x, \quad \text{if } x > 10.$$

Now $(dP/dx) = 0$ when $x = 115/3$. Hence the change in lease charge per car increases the optimal fleet size for maximum profit.

2.3 STOCK CONTROL

1. The annual cost of holding the filter in stock is given by

$$C = 5\left(\frac{365}{x}\right) + 20\,000 + \frac{0.18}{2}\left(\frac{10\,000x}{365}\right)2 \qquad (5.1)$$

$$\text{Cost of orders} \quad \text{Cost of filters} \quad \text{Stockholding}$$

where an order is placed every x days, so that $10\,000(x/365)$ is the size of each order.

Solutions to the exercises

The minimum cost is obtained when $(dC/dx) = 0$; i.e. when $x = 19.24$. Hence for a real solution; the filters are ordered every 19 days and to sell 10 000 per year; the order size is 521.

2. If ordering monthly, the size of the order is $10\,000/12 = 834$. If ordering fortnightly, the order size is 417. Using eqn (5.1), the difference in the annual costs is

   (a) For a monthly order £$(210.12 - 189.75) = $ £20.37

   (b) For a fortnightly order £$(195.06 - 189.75) = $ £5.31.

2.4 SALES RESPONSE TO ADVERTISING

1. The basic ordinary differential equation which models the sales response is

$$\frac{ds}{dt} = cA(t)\frac{(M-s)}{M} - bs.$$

For $A(t) = \alpha t$ and $b = 0$, we have

$$\frac{ds}{dt} = c\alpha t\frac{(M-s)}{M}.$$

Solving this equation, we have

$$\int \frac{1}{M-s}\,ds = \frac{c\alpha}{M}\int^t dt$$

so

$$-\ln(M-s) = \frac{c\alpha}{2M}t^2 + \text{constant}.$$

Hence

$$s = M - (M - S_0)e^{-c\alpha t^2/2M}.$$

The sales rate approaches the saturation level quite rapidly.

2.5 RADIOACTIVE DECAY

1. From the test we have for xenon 133, $k = (\ln 2)/5 = 0.139$. The amount of xenon 133 remaining after t days is

$$N(t) = N_0 e^{-kt}. \tag{5.2}$$

Therefore with $N_0 = 5$, $k = 0.139$, and $t = 10$,

$$N(10) = 1.25$$

and the weight of xenon 133 that has decayed in 10 days is

$$(5 - 1.25)\text{g} = 3.75\,\text{g}.$$

If we begin with N_0 g of xenon, then when 10 per cent has decayed we require the value of t for $N = (9/10)N_0$. From eqn (5.2) we obtain

$$t = -\frac{1}{k}\ln\left(\frac{9}{10}N_0/N_0\right) \simeq \tfrac{3}{4}\,\text{day}.$$

2. If $k = 2.77 \times 10^{-2}$, then the half-life for strontium 90 is given by

$$\tau = \frac{\ln 2}{k} \simeq 25\,\text{days}.$$

The time for 1 kg of strontium 90 to decay to 0.25 kg is

$$t = -\frac{1}{k}\ln\left(\frac{0.25}{1}\right) \simeq 50\,\text{days}.$$

113

Solutions to the exercises

3. According to the differential equation modelling the radioactive decay, the initial decay rate of 1 kg of uranium 238 is $kN(0)$ (from $(dN/dt) = -kN(t)$). Now

$$k = \frac{\ln 2}{\tau} = \frac{\ln 2}{4.5 \times 10^9} \quad \text{and} \quad N(0) = 1 \text{ kg.}$$

Hence the initial decay rate $= 1.54 \times 10^{-10}$ kg per year. At time t the decay rate is $kN(t)$ so that when the decay rate is 10^{-10} kg per year we have the amount of uranium 238

$$N(t) = \frac{10^{-10}}{k} \simeq 0.65 \text{ kg.}$$

2.6 CARBON DATING

1. With $\tau = 5568$ (in years) and $R_w = 6.68$, then for each sample we can estimate its life using the equation

$$t = \frac{5568}{\ln 2} \ln\left(\frac{6.68}{R}\right) \quad \text{(in years).}$$

So

 (i) $t = 281$ years;

 (ii) $t = 1135$ years;

 (iii) $t = 9219$ years.

2. If there is a 5 per cent possible error in the measurements, then the error in the estimation for the age of the 'Winchester round table' will be a maximum if

$$R_w = 6.68 + \left(\frac{5}{100} \cdot 6.68\right) = 7.014$$

and

$$R_c = 6.08 - \left(\frac{5}{100} \cdot 6.08\right) = 5.776.$$

Hence with these values we have an estimation for the age as

$$t = \frac{5568}{\ln 2} \ln\left(\frac{7.014}{5.776}\right) \simeq 1560 \text{ years.}$$

Thus this value places the construction date for the 'round table' at around the fifth century, and it could be King Arthur's table. This exercise illustrates the care needed in dating objects by 'carbon-dating'. Slight errors in the decay rates can change the age estimates quite considerably.

2.7. ART FORGERIES

1. If the paintings are sixteenth century then we may choose t_0 to be 400. Hence

$$kp(400) = kp(t)e^{400k} - R_0(e^{400k} - 1).$$

The values for $kp(400)$ for each of the five paintings are shown in the following table

	A	B	C	D	E
$kp(400)$	1.07×10^6	2.67	2.79×10^6	-27.53	8.33

Since $kp(400)$ should lie in the range 0–200 we might suspect paintings A, C, and D to be forgeries. (Take $k = 3.15 \times 10^{-2}$.)

2. Starting with $B = (kp(0) - R_0)e^{-kt_0} + R_0$, we can write $p(0)$ as a linear function of R_0. We get

$$p(0) = \frac{B}{k} e^{kt_0} + \frac{(1 - e^{kt_0})}{k} R_0.$$

Figure 5.5 shows a sketch of $p(0)$ against R_0 for $B = 200$ and $B = 0$. The shaded region indicates the allowable values for B.

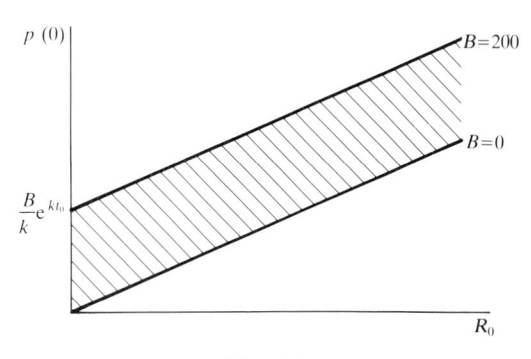

Fig. 5.5

2.8. DRUG ABSORPTION

1. If the half-time of elimination is 135 minutes, then

$$k = \frac{\ln 2}{135}$$

After 8 hours, the drug concentration becomes

$$y = 200 e^{-((\ln 2/135) \times 8 \times 60)}$$

$$= 17 \text{ mg}.$$

A dose of 183 mg is required to maintain a saturation level.

2. For aspirin, if the half-time of elimination is 45 minutes, we have $k = (\ln 2)/45$. The saturation level reached by a person taking y_0 mg every τ minutes is

$$y_s = \frac{y_0}{1 - e^{-k\tau}}.$$

With $y_0 = 1$ g and $\tau = 120$ minutes we have

$$y_s = \frac{1}{1 - e^{-((\ln 2/45) \times 120)}} = 1.19.$$

The saturation level is 1.19 g.

(i) According to the theory, if y_0 of caffeine is present in the blood at time $t = 0$, then at time t the concentration $y(t)$ is given by $y(t) = y_0 e^{-kt}$. If after 60 minutes (i.e. 1 hour), 25 per cent of the drug has been cleared, then

$$y(60) = \tfrac{3}{4} y_0, \quad \text{so that}$$

$$k = -\frac{1}{60} \ln(\tfrac{3}{4}) = 4.795 \times 10^{-3}.$$

Hence the half-time of elimination, τ, is

$$\tau = \frac{\ln 2}{k} = 144.6 \text{ minutes}.$$

(i.e. almost $2\tfrac{1}{2}$ hours).

(ii) If a person drinks 10 cups of coffee per day, then on average this is one cup every 2.4 hours and an intake of between 100 and 150 mg of caffeine every 2.4 hours (= 144 minutes).

Solutions to the exercises

The saturation level of caffeine will be between

$$\frac{100}{1-e^{-144k}} \text{ mg} \quad \text{and} \quad \frac{150}{1-e^{-144k}} \text{ mg},$$

i.e. between 200 and 300 mg (approximately).

(iii) The convulsive dose of caffeine is 10^4 mg (i.e. 10 g). Suppose that each cup of coffee contains 150 mg (i.e. we consider the 'worst' case) then to receive 10^4 mg of caffeine we must drink a cup of coffee every T minutes where T is given as a solution of the equation

$$10^4 = \frac{150}{1-e^{-kt}}$$

so that

$$T = -\frac{1}{k}\ln\left(1 - \frac{150}{10^4}\right) = 3.15 \text{ minutes}.$$

If we drink a cup of coffee every 3.15 minutes for a very long time (a lot more than 67 cups!!) we would take in a convulsive dose of caffeine. We would probably have made ourselves sick long before this by the shear volume of fluid!!

2.9. POPULATION MODELS

1. The growth rate γ is 6/1000 per year and since it is positive the population increases. Using eqn (2.15) the population in year t is given by

$$N(t) = N_0 e^{\gamma t}$$

where $N_0 = 3000$ in 1980 ($t = 0$) and $\gamma = 0.006$.

(i) In 1985, $N = 3000 e^{0.006 \times 5} = 3091$;

(ii) In 1990, $N = 3000 e^{0.006 \times 10} = 3185$;

(iii) In 2000, $N = 3000 e^{0.006 \times 20} = 3382$;

(where each answer is rounded down to a whole number).

2. (a) The equation describing the population change becomes

$$\delta N = \alpha N \delta t - \beta N \delta t + 100 \delta t.$$

Dividing by δt and letting $\delta t \to 0$ we get

$$\frac{dN}{dt} = (\alpha - \beta)N + 100.$$

(b) The differential equation in (a) has solution

$$N(t) = -\frac{100}{(\alpha - \beta)} + \left(N_0 + \frac{100}{(\alpha - \beta)}\right)e^{(\alpha-\beta)t}.$$

(c) If $\alpha = \beta$, we would expect the population to increase at a rate of 100 individuals per unit time. The solution in (b) does not apply in this case because of the singularity in the term $-100/(\alpha - \beta)$. Instead if we go back to the differential equation in (a), for $\alpha = \beta$ we have

$$\frac{dN}{dt} = 100 \quad \text{and then}$$

$$N = 100t + N_0$$

which agrees with our expectation.

3. The population in exercise 1 is given by

$$N(t) = 3000 e^{0.006 t}.$$

The population will double after τ years, when $N = 6000$, where

Solutions to the exercises

$$\tau = \frac{1}{0.006}\ln 2 = 115.5,$$

i.e. in the year 2095.

4. Plotting the growth rate curve, we get Fig. 5.6. We then have $a = 0.414$ and $N_\infty = 56.17 \times 10^6$. With $N_0 = 55.61 \times 10^6$ in 1971, the predicted population is

Year	1971	1972	1973	1974	1975	1976	1977	1978	1979
Predicted population ($\times 10^6$)	55.61	55.80	55.92	56.01	56.06	56.10	56.1	56.1	56.1
Actual population ($\times 10^6$)	55.61	55.78	55.91	55.92	55.90	55.89	55.85	55.84	55.88

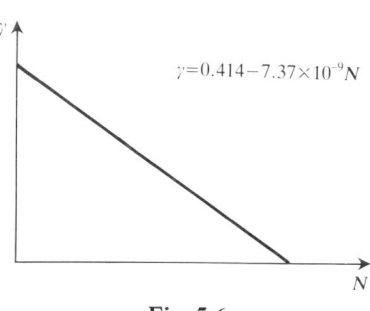

$y = 0.414 - 7.37 \times 10^{-9} N$

Fig. 5.6

5.
$$A = \begin{bmatrix} 0.915 & 0.075 & 0 \\ 0.065 & 0.959 & 0 \\ 0 & 0.033 & 0.938 \end{bmatrix}$$

$$A^2 = \begin{bmatrix} 0.842 & 0.141 & 0 \\ 0.122 & 0.925 & 0 \\ 0.00215 & 0.0626 & 0.880 \end{bmatrix}$$

$$A^4 = (A^2)^2 = \begin{bmatrix} 0.726 & 0.248 & 0 \\ 0.215 & 0.872 & 0 \\ 0.0113 & 0.113 & 0.774 \end{bmatrix}$$

$$A^8 = (A^4)^2 = \begin{bmatrix} 0.581 & 0.397 & 0 \\ 0.334 & 0.814 & 0 \\ 0.0413 & 0.189 & 0.599 \end{bmatrix}$$

$$A^{10} = A^8 A^2 = \begin{bmatrix} 0.537 & 0.449 & 0 \\ 0.389 & 0.801 & 0 \\ 0.0592 & 0.218 & 0.527 \end{bmatrix}$$

The 1891 population for the UK is then estimated using

$$N(10) = A(10)N(0) = \begin{bmatrix} 10\,327 \\ 13\,019 \\ 5\,686 \end{bmatrix} \quad \text{(in thousands)}$$

i.e. $29\,032 \times 10^3$ people. The actual 1891 population was $37\,600\,000$.

6. From the data given we can write the initial population for the three age groups, under 15, 15–44, and over 45, in matrix form as

Solutions to the exercises

$$N(0) = \begin{bmatrix} 13.1 \\ 22.1 \\ 20.9 \end{bmatrix}.$$

In the year 2000, 24 years later we have

$$N(24) = A^{24} N(0)$$

With

$$A = \begin{bmatrix} 0.915 & 0.075 & 0 \\ 0.065 & 0.959 & 0 \\ 0 & 0.033 & 0.938 \end{bmatrix},$$

$$A^{24} = \begin{bmatrix} 0.466 & 0.638 & 0 \\ 0.553 & 0.840 & 0 \\ 0.179 & 0.344 & 0.215 \end{bmatrix},$$

and

$$N(24) = \begin{bmatrix} 20.20 \\ 25.81 \\ 14.44 \end{bmatrix},$$

the predicted population for the year 2000 is 60.45 million.

Solutions to the exercises

3. Mechanics

3.1. PURSUIT CURVES

1. With $a = 2500, b = 3000, v_A = 1000, v_m = 2000$ we have
$$t = \frac{(2500^2 + 3000^2)^{1/2} \div 2500 \times k}{200 \times (1-k^2)}$$
where
$$k = \frac{v_A}{v_m} = \frac{1000}{2000} = \frac{1}{2}.$$
Hence $t = 3.44$ seconds.

3. The initial positions of the dog and cat, and at time t are shown in Fig. 5.7. At time t the position of the cat is
$$x_C = vt, \quad y_C = 0.$$

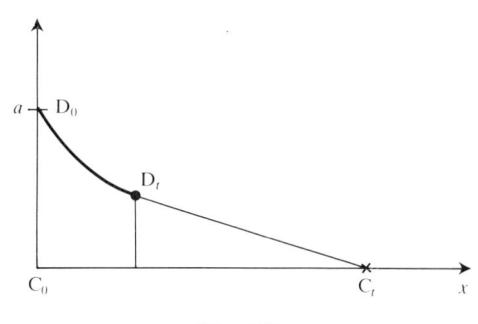

Fig. 5.7

If the position of the dog is (x_D, y_D), then
$$\left(\frac{dy}{dx}\right)_D = \frac{-y_D}{x_C - x_D} = \frac{-y_D}{vt - x_D}$$
and $v_D^2 = \dot{x}_D^2 + \dot{y}_D^2 = w^2$. The path of the dog satisfies the differential equation,
$$-y\frac{dx}{dy} = vt - x.$$
Following the same method as the text, differentiate with respect to t
$$-\dot{y}\frac{dx}{dy} - y\frac{d^2x}{dy^2}\dot{y} = v - \dot{x},$$
i.e.
$$-y\dot{y}\frac{d^2x}{dy^2} = v \tag{5.3}$$
since $\dot{y}(dx/dy) = \dot{x}$. Now $\dot{x}^2 + \dot{y}^2 = w^2$ so that
$$\dot{y}^2 = \frac{w^2}{(1 + \dot{x}^2/\dot{y}^2)} = \frac{w^2}{[1 + (dx/dy)^2]}.$$
Substituting into (5.3) gives
$$\frac{d^2x}{dy^2} = -\frac{[1 + (dx/dy)^2]^{\frac{1}{2}}}{y}\left(\frac{v}{w}\right).$$

119

Solutions to the exercises

Let $p = (dx/dy)$ and $k = (v/w)$. Then

$$\frac{dp}{dy} = \frac{-k(1+p^2)^{1/2}}{y}.$$

Integrating

$$\ln[p + (1+p^2)^{1/2}] = -k \ln y + A$$

where A is constant. When $t = 0$, $y = a$ and

$$p = \frac{1}{(dy/dx)} = \frac{1}{\infty} = 0.$$

With these values $A = k \log_e a$. Thus

$$p + (1+p^2)^{1/2} = \left(\frac{a}{y}\right)^k$$

or

$$(1+p^2)^{1/2} = \left(\frac{a}{y}\right)^k - p.$$

Squaring and then solving for p we get

$$p = \frac{dx}{dy} = \frac{\left(\frac{a}{y}\right)^{2k} - 1}{2\left(\frac{a}{y}\right)^k}.$$

Integrating with respect to y gives the equation of the path of the dog as

$$x = \frac{1}{2}\left[\frac{a^k y^{-k+1}}{1-k} - \frac{y^{k-1}}{(k+1)a^k}\right] + B$$

where B is constant and $k \neq 1$. When $x = 0$, $y = a$ so that

$$B = \frac{1}{2}\left[\frac{a}{k+1} - \frac{a}{1-k}\right] = \frac{-ka}{1-k^2}.$$

If $k = 1$, then

$$x = \frac{1}{2}\left[a \ln y - \frac{y^2}{2a}\right] + B'$$

With $x = 0$, $y = a$ we have

$$B' = -\frac{1}{2}\left[a \ln a - \frac{a}{2}\right].$$

The dog catches the cat when $y = 0$.

Consider each case $k < 1$, $k = 1$, and $k > 1$ in the equation

$$f(y) = \left[\frac{a^k}{y^{k-1}} - \frac{y^{k+1}}{(k+1)a^k}\right] - \frac{2ka}{1-k^2} = 0.$$

These cases correspond to the dog's speed being greater than ($k < 1$), equal to ($k = 1$), and less than ($k > 1$) the cat's speed. Figure 5.8 shows sketches of $f(y)$ for the three cases. We see that the dog can catch the cat provided its speed is greater than or equal to the cat's speed. A result that is perhaps intuitively obvious.

3.2. MODELLING RIVER FLOW

1. If $\mathbf{v_R}$ is the velocity of the river then the velocity of the boat is $\mathbf{v} = \mathbf{v_R} + \mathbf{i}$ (choosing $\mathbf{i}$ perpendicular to the bank). Now $\mathbf{v_R} = v_1 \mathbf{j}$ (there is no flow from bank to bank) hence

$$\mathbf{v} = \mathbf{i} + v_1 \mathbf{j}$$

Solutions to the exercises

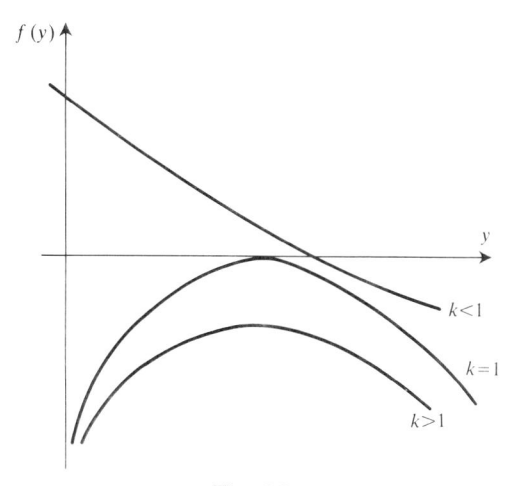

Fig. 5.8

so that the path is given as a solution of

$$\frac{dx}{dt} = 1 \quad \text{and} \quad \frac{dy}{dt} = v_1.$$

Hence $x = t$ for each model.

Model 1.

For $0 \leqslant x \leqslant 10$,

$$\frac{dy}{dt} = 0.3x = 0.3t.$$

Integrating,

$$y = \frac{0.3}{2} t^2.$$

For $10 \leqslant x \leqslant 20$,

$$\frac{dy}{dt} = -0.3x + 6 = -0.3t + 6$$

$$y = -\frac{0.3}{2} t^2 + 6t - 30.$$

Eliminating t, we have the path

$$y = \frac{0.3}{2} x^2, \quad 0 \leqslant x \leqslant 10$$

$$y = -\frac{0.3}{2} x^2 + 6x - 30, \quad 10 \leqslant x \leqslant 20.$$

The boat reaches the far bank when $x = 20$, i.e. a distance 30 m downstream.

Model 2.

$v_1 = 0.03x(20 - x)$. Hence,

$$\frac{dy}{dt} = 0.03x(20 - x) = 0.03t(20 - t).$$

Integrating,

$$y = 0.3t^2 - 0.01t^3.$$

Eliminating t,

Solutions to the exercises

$$y = 0.3x^2 - 0.01x^3.$$

The boat travels a distance 40 m downstream.

Model 3

$$v_1 = 3 \sin \frac{\pi}{20} x$$

Hence

$$\frac{dy}{dt} = 3 \sin \frac{\pi}{20} x = 3 \sin \frac{\pi}{20} t.$$

Integrating,

$$y = -\frac{60}{\pi} \cos \frac{\pi}{20} t + \frac{60}{\pi}.$$

Eliminating t,

$$y = \frac{60}{\pi} - \frac{60}{\pi} \cos \frac{\pi}{20} x.$$

The boat travels a distance 38.2 m downstream. Clearly Model 2 gives furthest distance downstream.

2. Similar models to those in the text can be formulated. For example, we could write Model 1 as

$$v = \frac{4}{15} x, \qquad 0 \leq x \leq 15$$

$$= -\frac{4}{5} x + 16 \quad 15 \leq x \leq 20.$$

and Model 2 as

$$v = \frac{4}{75} x(20 - x).$$

3. The highest speed occurs where $r = 0$ and v is 2 m s^{-1}. The rate of flow is

$$\int_0^{10} v(\pi r\,dr) = \int_0^{10} \pi \frac{(10-r)^2}{20} r\,dr = 52.4 \text{ m}^3 \text{ s}^{-1}.$$

The area of the cross-section of the river is $\frac{1}{2}(\pi \times 10^2)$. Thus a constant speed of $\frac{1}{3} \text{m s}^{-1}$ would give the required flow rate.

3.3. THE TENNIS SERVICE

1. The minimum height is such that $x = 21$ (in feet). From Fig. 3.13 (p. 71) the angle DPA is given by arc $\tan(3/21)$. The the minimum height $h = 60 \tan \alpha = 8.57$ (in feet).

2. An improvement to the model might be to consider the ball as a particle that clears the net at roughly half a diameter above the net and lands roughly half a diameter within the service lines.

3.(i) For $\alpha = 9$ and $h = g$ the top of the net is *just* on the path of the ball if the point $x = 39, y = -6$ satisfies the equation of the path. Thus

$$-6 = -\frac{g \times 39^2}{2v^2}.$$

Hence

$$v = \left(\frac{39^2 \times g}{12}\right)^{1/2} = 35.2 \text{ feet per second.}$$

(ii) For the ball to strike the court at B, the point $x = 60, y = -9$ satisfies the equation of the path. Thus

Solutions to the exercises

$$-9 = -\frac{g \times 60^2}{2v^2}$$

Hence

$$v = \left(\frac{60^2 \times g}{18}\right)^{1/2} = 44.3 \text{ feet per second.}$$

4.   If $v = 220$, the top of the net lies on the path if

$$-6 = 39 \tan \alpha - \frac{39^2 \times g}{2 \times 220^2} \sec^2 \alpha,$$

i.e.

$$-6 = 39 \tan \alpha - 0.154 \sec^2 \alpha.$$

Putting $\sec^2 \alpha = 1 + \tan^2 \alpha$, then $\alpha = -8.52°$, i.e. the player serves at an angle of $8.52°$ below the horizontal.

The ball hits the court a distance x from A where x is a solution of the quadratic equation

$$\frac{gx^2}{2v^2} \sec^2 \alpha - x \tan \alpha + 9 = 0.$$

With $\alpha = 8.52$ and $v = 220$, we get $x = 62.8$. This point is just beyond the service line so that the service is just out of court. The first model would give a distance $39 + 117/(h \cdot 3)$ from the server (using Fig. 3.13, p. 000). With $h = 9$ this is 58.5 feet, i.e. just within the service line. For this model $\alpha = 8.75°$.

3.5. BRAKING A CAR

1.	Car speed (m.p.h.)	10	20	30	40	50	60	70
	Stopping distance (feet)	15	40	75	120	175	240	315

2. $d_b(\text{ft}) = 0.305 d_b(\text{m}) = d'_b(\text{m})$

$v(\text{m.p.h.}) = 1.6 v(\text{km/h}) = v'(\text{km/h})$

So using $d_b = 1/20 v^2$, we need to find d'_b in terms of v'. Now

$$d_b = \frac{d'_b}{0.305} \quad \text{and} \quad v = \frac{v'}{1.61}.$$

Therefore

$$\frac{d'_b}{0.305} = \frac{1}{20} \times \left(\frac{v'}{1.61}\right)^2, \text{ i.e.}$$

$$d'_b = \frac{0.305}{20 \times 1.61^2} v'^2, \text{ i.e.}$$

$$d'_b = 0.0059 v'^2 = \frac{v'^2}{170}$$

Thus $d_b \simeq (1/170) v^2$ if d_b is in m and v in km/h. For example, for $v = 48$ km/h (30 m.p.h.),

$$d_b = \frac{1}{170} \times 48^2 = 13.6 \text{ m.}$$

3. (a) and (b)

If $F = pmg$, the constant acceleration is $-pg$. So, using $v^2 - u^2 = 2$ as with $v = 0, u = 30, a = -pg$, and $s = d_b$, then $-900 = -2pg d_b$, i.e.

$$d_b = \frac{450}{pg} \simeq \frac{14}{p} \quad \text{with} \quad g = 32.2 \text{ ft/s}^2.$$

123

Solutions to the exercises

This gives

p	0	0.2	0.4	0.6	0.8	1.0	1.2	1.4	1.6	1.8	2.0
d_b(feet)	∞	70	35	23	17.5	14	11.7	10	8.8	7.8	7.0

(c) Following the above argument but with $u = v$, instead of 30 gives

$$-v^2 = -2pgd_b, \text{ i.e.}$$

$$d_b = \frac{-v^2}{2pg} = \frac{v^2}{64.4p}$$

4.
$$v\frac{dv}{dx} = -(0.660 - 0.001v)g.$$

Therefore,

$$\int_v^0 \frac{-v}{0.660 - 0.001v}\,dv = \int_0^{d_b} g\,dx$$

$$\int_v^0 1000\left(\frac{0.660 - 0.001v - 0.660}{0.660 - 0.001v}\right)dv = \int_0^{d_b} g\,dx$$

$$\int_v^0 \left(1000 - \frac{660 \times 1000}{660 - v}\right)dv = \int_0^{d_b} g\,dx$$

$$[1000v + 660 \times 1000 \ln(660 - v)]_v^0 = [gx]_0^{d_b}$$

$$660 \times 1000 \ln|660| - 1000v - 660 \ln|660 - v| = gd_b$$

$$660 \times 1000 \ln\left|\frac{660}{660 - v}\right| - 100v = gd_b,$$

i.e.

$$d_b = \frac{1000}{g}\left(660 \ln\left|\frac{660}{660 - v}\right| - v\right).$$

Once again, if v is measured in m.p.h. not ft/s we replace v by $1.47v$. Then

$$d_b = \frac{1000}{g}\left(660 \ln\left|\frac{660}{660 - 1.47v}\right| - 1.47v\right)$$

$$= 47.9, \quad \text{when } v = 30 \text{ (compared with 45)}$$

$$= 201.2, \quad \text{when } v = 60 \text{ (compared with 180)}.$$

3.7. KEPLER'S LAW

1. Comparing Bode's law with the data of Table 3.4, we have

Planet	n	R/R_{earth}	
		Bode's law	Observation
Mercury	$-\infty$	0.4	0.387
Venus	0	0.7	0.723
Earth	1	1	1
Mars	2	1.6	1.52
Jupiter	4	5.2	5.2
Saturn	5	10	9.54
Uranus	6	19.6	19.18
Neptune	7	38.8	30.06
Pluto	8	77.2	39.49

Solutions to the exercises

Bode's law provides a reasonable model for all planets except for Neptune and Pluto. It is interesting to note that Pluto would fit $n = 7$ in Bode's law, thus giving some credence to the theory that Pluto was a moon of Neptune until some catastrophe parted them.

2. If the gravitational law were an inverse cube law, then

$$\dot{R}\theta^2 = \frac{\gamma M}{R^{3'}}$$

so that

$$\dot{\theta} = \frac{(\gamma M)^{1/2}}{R^2}.$$

Integrating and rearranging as before,

$$T = \frac{2\pi R^2}{(\gamma M)^{1/2}}$$

and Kepler's third law would be changed.

3.8. INDUSTRIAL LOCATION

2. *Three customers.* If $w_1 = w_2 = w_3$, the solution is relatively straightforward. Provided the triangle formed by three customers has all its angles less than $120°$, the depot will be located such that the lines to the customer all make angles of $120°$. If the triangle has an angle rather than $120°$ then the depot will be located at that position.

Each customer is 'pulling' with the same 'force'; hence, if the depot is at D, the forces D1, D2, and D3 are equal. Therefore the triangle of forces is equilateral and the three forces make angles of $120°$ with each other.

If one of the angles of the triangle is greater than or equal to $120°$, it is not possible for the forces to be at $120°$ to each other; hence the optimum is to put the depot at this angle as it comes as close as possible to the ideal solution (Fig. 5.9).

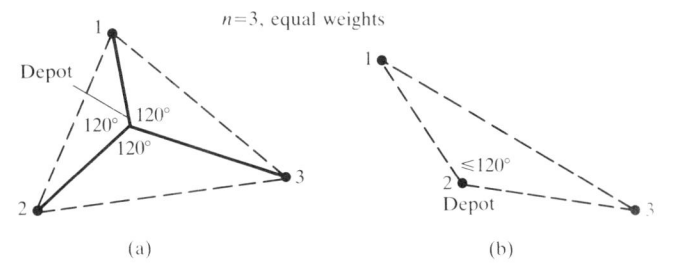

Fig. 5.9

Solutions to the exercises

4. Probability and statistics

4.1. ARE YOU BEING SERVED?

1. The null hypothesis is that the coin is unbiased. If this is true, the probability of throwing a head on a single toss of the coin is 0.5, i.e. $p = 0.5$.

    The mean of the distribution $= np = 200(0.5) = 100$.

    The standard deviation of the distribution $= \sqrt{(npq)^{\frac{1}{2}}} = \{200(0.1)(0.9)\}^{\frac{1}{2}} = 7.07$

    The normal approximation to the binomial distribution is valid since $np > 5$ and $p > 0.1$.

    Standardizing the sample data (i.e. 80) gives z,

    $$z = \frac{80 - 100}{7.07} = -2.83.$$

Using the normal distribution table, we see that a measurement of -2.83 or less will occur about 0.23 per cent of the time (Fig. 5.10).

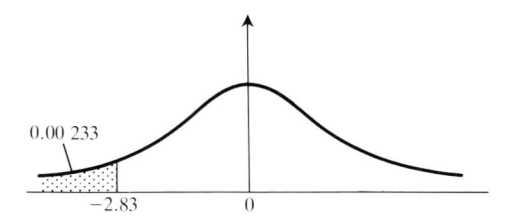

Fig. 5.10

If the coin is fair, then 80 or fewer heads out of 200 tosses would be expected only 0.23 per cent of the time. This suggests that the coin is biased in favour of tails.

2. On a single toss of a fair coin, the probability of obtaining a head = probability of obtaining a tail = 0.5. Suppose a coin is tossed n times. Then the probability of obtaining n tails is $(0.5)^n$, i.e. the probability of obtaining no heads $= (0.5)^n$.

    We require the probability of obtaining no heads to be 5 per cent or less

    $$(0.5)^n \leqslant 0.05$$

    $$n \log(0.5) \leqslant \log(0.05)$$

    $$n \geqslant \frac{\log(0.05)}{\log(0.5)} = 4.32$$

    $n = 5$ (since n must be an integer)

    Note that $(0.5)^5 = 0.03$.

    If a coin is tossed five times, there is a 97 per cent chance of obtaining at least one head.

3. The probability of throwing a 6 on a single toss of a die is $\frac{1}{6}$. If the die is thrown n times, the probability that no 6s occur is $(\frac{5}{6})^n$.

    We require $(\frac{5}{6})^n \leqslant 0.25$

    $$n \geqslant \frac{\log(0.25)}{\log(\frac{5}{6})} = 7.6$$

    $$n = 8.$$

126

Solutions to the exercises

Note that $(\frac{5}{6})^8 = 0.23$, so if a die is thrown 8 times, there is a 77 per cent chance of at least one 6 occurring.

4. The probability a coach breaks down = 0.1.

   (i) The probability that all coaches are in service (i.e. no breakdowns) = $(0.9)^{20} = 0.12$.

   (ii) The probability that three coaches are out of service is

   $$^{20}C_3(0.1)^3(0.9)^{17} = \frac{20!}{17!\,3!}(0.1)^3(0.9)^{17} = 0.19.$$

   (iii) The probability that more than three coaches are out of service is

   $1 - $ [probability that three or fewer are out of service]

   $= 1 - {^{20}C_0}(0.1)^0(0.9)^{20} - {^{20}C_1}(0.1)^1(0.9)^{19} - {^{20}C_2}(0.1)^2(0.9)^{18} = {^{20}C_3}(0.1)^3(0.9)^{17}$

   $= 0.13.$

4.2. QUALITY CONTROL

1. Let x = torque $-$ breaking down strength. Then a cap will break if $x > 0$. The variable x is normally distributed

$$\text{Mean} = 8 - 12 = -4.$$

Standard deviation = $\sqrt{\{(1.2)^2 + (1.6)^2\}} = 2$.

Standardizing 0 gives

$$\frac{0-(-4)}{2} = 2.$$

From normal tables, the probability of the cap breaking is 0.0228 (Fig. 5.11), i.e. just over 2 per cent of caps will be broken.

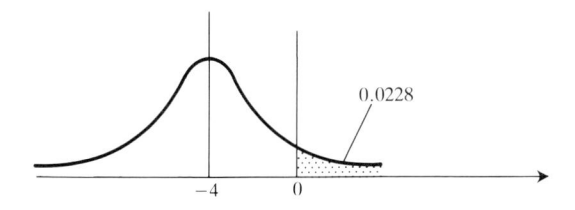

Fig. 5.11

2. (a) The weight of four adults is normally distributed.

$$\text{Mean} = 4 \times 75 \text{ kg} = 300 \text{ kg}.$$

$$\text{Standard deviation} = \sqrt{4} \times 15 \text{ kg} = 30 \text{ kg}.$$

Standardizing 350 gives

$$\frac{350 - 300}{30} = 1.67.$$

The probability that the weight of four adults is less than 350 kg is 0.9525 (Fig. 5.12). The probability that the plane can take off safely is 95.25 per cent.

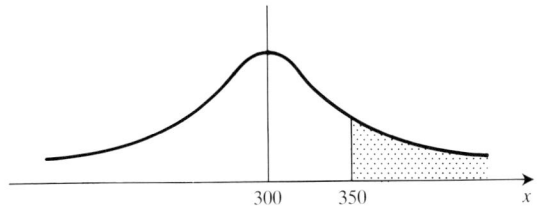

Fig. 5.12

Solutions to the exercises

(b) The weight of four adults and one child is normally distributed.

$$\text{Mean} = (4 \times 75) + 23 \text{ kg} = 323 \text{ kg}.$$

$$\text{Standard deviation} = \sqrt{\{(4 \times 225) + 1 \times 49\}} \text{ kg}$$

$$= 30.8 \text{ kg}.$$

Standardizing 350 gives

$$\frac{350 - 323}{30.8} = 0.88.$$

The probability that the weight is less than 350 kg is 0.8016. The probability that the plane can take off safely is 80.16 per cent.

If 40 kg of luggage are included in (a), the mean is now 340 kg. The standard deviation remains unchanged.

Standardizing 350 gives

$$\frac{350 - 340}{30} = 0.32.$$

The probability that the combined weight is less than 350 kg is 0.6293. The probability that the plane can take off safely is 62.93 per cent.

If 40 kg of luggage are included in (b) the mean weight is now 363 kg. The standard deviation remains unchanged at 30.8 kg.

Standardizing 350 gives

$$\frac{350 - 363}{30.8} = -0.42.$$

The probability that the weight is greater than 350 kg is 0.6628. So the probability that the weight is less than 350 kg is 0.3372. The probability that the plane can take off safely is 33.72 per cent.

3. For a normal distribution, 95 per cent of measurements lie within mean ± 1.96 standard deviations. (Fig. 5.13).

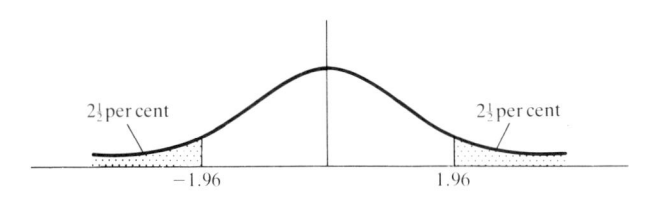

Fig. 5.13

Thus for the rods, 1.96 standard deviations = 0.2 mm. Hence standard deviations = 0.2/1.96 mm.

The combined length of the two rods is normally distributed

$$\text{Mean} = 50 \text{ mm} + 20 \text{ mm} = 70 \text{ mm}.$$

$$\text{Standard deviation} = \sqrt{2} \frac{0.2}{1.96}.$$

Hence, 95 per cent tolerances for the combined length are

$$70 \pm 1.96 \left(\sqrt{2} \frac{0.2}{1.96} \right) \text{mm}$$

$$= 70 \pm (\sqrt{2}) 0.2 \text{ mm}$$

$$= 70 \pm 0.28 \text{ mm}.$$

4. The time for each call is normally distributed

Solutions to the exercises

(i)
$$\text{Mean} = 30 \text{ min.}$$
$$\text{Standard deviation} = 6 \text{ min.}$$

The time for 15 calls is normally distributed
$$\text{Mean} = 15 \times 30 \text{ min} = 450 \text{ min.}$$
$$\text{Standard deviation} = \sqrt{15} \times 6 \text{ min} = 23 \text{ min.}$$

Standardizing 480 min (8 hours) gives
$$\frac{480 - 450}{23} = 1.30.$$

The probability that the time is greater than 480 minutes is 0.0968 (Fig. 5.14), i.e. there is just under a 10 per cent chance the salesman will have to work overtime.

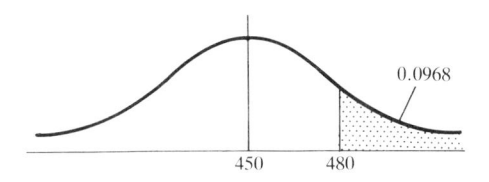

Fig. 5.14

(ii) Let $f = 480 - $ (time to make 15 calls), i.e. f is the salesman's free time per day (in minutes). f is normally distributed.

$$\text{Mean} = 30 \text{ min.}$$
$$\text{Standard deviation} = 23 \text{ min.}$$

The 'free' time per week is normally distributed.

$$\text{Mean} = 150 \text{ min.}$$
$$\text{Standard deviation} = \sqrt{5} \times 23 \text{ min} = 51 \text{ min.}$$

To leave a 'tail-end' of $\frac{1}{2}$ per cent we must be 2.58 standard deviations from the mean (Fig. 5.15). Hence, the 90 per cent confidence interval for free time is

$$150 \pm (2.58 \times 51) \text{ min}$$
$$= 18.4 \text{ to } 281.6 \text{ min}$$

i.e. there is a 99 per cent chance the salesman's free time per week is between 18 and 282 min.

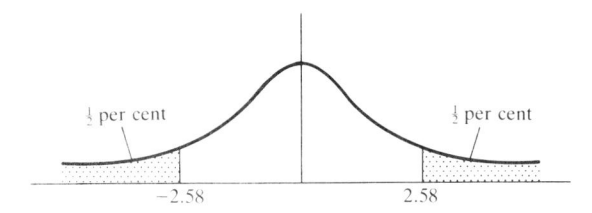

Fig. 5.15

4.3. BLOOD DONORS

1. P (individual selected at random can receive blood type A+). Blood types able to receive A+ are those blood types that contain A and Rh, i.e. AB+ and A+. Therefore,

$$P(\text{suitable recipient for A+}) = P(\text{AB+}) + P(\text{A+})$$
$$= 0.34 + 0.357$$
$$= 0.391.$$

129

Solutions to the exercises

2.

Blood type	Probability of occurrence	Suitable recipient	Probability that random selection produces suitable recipient
O−	0.066	O−, A−, B−, AB−, O+, A+, B+, AB+	1.00
A−	0.063	AB−, A+, AB+, A−	0.46
B−	0.015	AB−, B+, AB+, B−	0.14
AB−	0.006	AB+, AB−	0.04
O+	0.374	A+, B+, AB+, O+	0.85
A+	0.357	AB+, A+	0.391
B+	0.085	AB+, B+	0.119
AB+	0.034	AB+	0.034

3. Blood type O is suitable for any other group and itself. Blood types AB+ are not suitable to be given to any types other than their own.

Blood types AB− are also suitable to be given only to a small proportion of recipients, themselves and AB+ with low probability of 0.04 in a random sample.

Blood type A− is suitable for almost half of a sample chosen at random.